PLAYING WITH FEELINGS

Playing with Feelings

Video Games and Affect

Aubrey Anable

University of Minnesota Press
Minneapolis
London

Portions of chapter 3 were published in a different form as "Casual Games, Time Management, and the Work of Affect," *Ada: A Journal of Gender, New Media, and Technology*, no. 2 (2013), http://adanewmedia.org; and as "Labor/Leisure," in *Time: A Vocabulary of the Present*, ed. Amy Elias and Joel Burges (New York: New York University Press, 2016). Reprinted by permission.

Published by the University of Minnesota Press
111 Third Avenue South, Suite 290
Minneapolis, MN 55401-2520
http://www.upress.umn.edu

ISBN 978-1-5179-0024-3 (hc)
ISBN 978-1-5179-0025-0 (pb)

A Cataloging-in-Publication record for this book is available from the Library of Congress.

The University of Minnesota is an equal-opportunity educator and employer.

UMP LSI

Contents

Introduction

Video Games as Structures of Feeling

I N 1983, the new video game company Electronic Arts (EA) introduced itself with a provocative advertising campaign that asked, "Can a computer make you cry?" The text that ran alongside a black-and-white photograph of the "software artists" read in part:

> What are the touchstones of our emotions?
> Until now, the people who asked such questions . . . were, in the traditional sense, artists.
> We're about to change that tradition. . . .
> In short, we are finding that the computer can be more than just a processor of data.
> It is . . . an interactive tool that can bring people's thoughts and feelings closer together. . . . And while fifty years from now, its creation may seem no more important than the advent of motion pictures or television, there is a chance it will mean something more.

Emerging in the context of a market oversaturated with poor-quality Atari games, EA hitched its video games to the "something more" promise of art.[1] This promise was premised on two related ideas: that art is defined by how it moves people emotionally and that computer-mediated interactivity can "bring people's thoughts and feelings closer" in ways that other art forms cannot. Now, thirty-some years out from this advertising campaign, its promise seems protracted. There is still an assumption in the popular discourse about video games that the main obstacle keeping them from attaining the cultural status of film and other arts is their limited capacity to affect our feelings. There is also a related assumption that this is changing now that games (finally!) are becoming more emotionally complex. The ability to evoke strong empathetic responses, like crying, continues to be posited as the litmus test for taking video games seriously. But this is an extremely narrow emotional criterion for any cultural form. Plenty of video games make plenty of people cry, yet we might hesitate before labeling all such games "art." Similarly, many games evoke nothing like tears but are regularly exhibited at major institutions of high art. Furthermore, in the history of the legitimation of film, for example, the capacity to arouse strong emotional responses has been negatively linked to the status of "art."[2]

I begin with this point about emotions, art, and video games not because I am interested in the issues of cultural legitimation that the word *art* performs but rather because of my interest in the way this prevalent discourse leaves gaming culture with an impoverished emotional and critical language, one that looks for big emotions in a medium that seems to traffic more interestingly in the minor affects. Video games have always had a tremendous capacity to affect our emotions, but they do so in ways that have important differences from the ways films or novels affect us. We might feel quite tense playing a first-person shooter, or, conversely, like Francis Underwood in the American version of *House of Cards,* we might find relaxation and cathartic release in the genre's repetition. Some first-person shooters are just boring, but boredom, as Siegfried Kracauer argues, is also an interesting and culturally significant feeling.[3]

A woman standing on a subway platform consults the screen above her. It displays the news headlines, weather forecast, and estimated arrival time of the next train. Her glance returns to her phone. Its networked functions constrained underground, she opens the game application *Candy Crush Saga* to pass the time. Stuck on level 65 for weeks now, she sighs and taps the screen over the "replay level" icon, feeling pretty sure that she will fail again. But the train is a couple minutes away, and, after that, her destination another ten minutes; the puzzle game offers a familiar way to pass time in transit. The grid of colorful candy-shaped icons appears, and the woman begins clearing the rows by matching the icons in threes and fours according to the metrics established by the level. Holding the phone in the palm of her hand, thumb hovering over the screen, she gently swipes the icons into place. With each match, the images burst and transform, and a new arrangement comes into view. The train arrives and the woman closes the game to board. Once seated, she opens the game and resumes where she left off. Occasionally she glances around, noticing others engaged with

In 1983, Electronic Arts proposed that video games would become the emotional touchstones of the future and that computers would "bring people's thoughts and feelings closer together."

their phones, many playing games, many, in fact, playing the same game that she is. This is entirely ordinary. She notes how many stops are left before her own and returns her attention to the game. She is running out of moves. This is frustrating, but only mildly so. Approaching her stop with no chance of winning, she hastily and disinterestedly matches a couple more pieces and closes the game. She failed the level again, and she has arrived at her destination. She tucks the phone into her pocket, and she does not think about the game again until her next commute.

This book is about this moment and moments similar to it. I have held this moment in my mind while writing each chapter, and I have held it in conversation with another moment in which a cultural theorist used the time and space of the commute to address the tones and textures of everyday life. In Raymond Williams's 1958 essay "Culture Is Ordinary," the first line, "The bus stop was outside the Cathedral," draws together the ordinary and that which transcends it.[4] From the bus stop near the Welsh village in which he grew up and through his subsequent ride, Williams takes the reader through the lessons of culture both as subjectively lived—a "whole way of life"—and as it circulates through film, literature, and television.[5] Williams makes a case for the ordinariness of culture, asserting that culture emerges neither from highbrow defenses of "true art" nor from working-class rejections of elitism but rather from everyday experiences. Williams sees culture in the Norman castles, blast furnaces, and movie posters that make up the landscape framed by the bus's windows, and he also sees culture in the bus driver and conductress, who are "deeply absorbed in each other," and in the tears his grandfather wept when he told the story of being forced out of his cottage.[6] Culture is material, locatable, but also ephemeral and subjectively experienced. Williams writes: "The questions I ask about our culture are questions about our general and common purposes, yet also questions about deep personal meanings. Culture is ordinary, in every society and in every mind."[7]

What if Williams had been playing *Candy Crush Saga* on his bus ride? The question seems irreverent, but it is one that I find worth asking when I play games on my phone in transit and watch others do the same. In these ordinary moments, what feelings are being mediated and playfully expressed through these devices, software, and images? During the same year that Williams wrote his essay, the Manhattan Project physicist William Higinbotham designed what many consider to be the first video game, *Tennis for Two,* at Brookhaven National Laboratory. As Patrick Jagoda argues, video games trace "the phenomenal residue of a historical convergence between technoscientific and aesthetic horizons" that began in the games' emergence from the military–industrial complex and Cold War.[8] This small media fragment, within the larger historical context of the atomic bomb and the military–industrial complex, seems banal. It is only in retrospect—with our present knowledge of the historical trajectory of computers and video games over the rest of the twentieth century—that this simple metaphorization of abstract computer processes as a friendly game of tennis played across two human bodies and a computer seems significant at all. It is also only in retrospect that *Tennis for Two* can even be read as a "video game." In 1958, while Williams pondered the shifting landscape of postwar Britain through a bus window, *Tennis for Two* was just a demonstration of computer processing for visitors' day at the lab in postwar America. In 1983, when the EA advertisement pointed fifty years into the future to a time when video games might be key emotional and cultural touchstones, video games had already been doing the affective labor of computers for a couple of decades. That EA could propose that video games would bring our "thoughts and feelings closer together" was at least in part due to the way research into artificial intelligence, cybernetics, and related intellectual developments from the mid-twentieth century had already used computer models to rethink the relationship between thoughts and feelings in human (and machine) consciousness.[9]

In this book, I put video games into conversation with affect theory and position "video games" and "affect theory" as related historical and intellectual projects, rather than self-evident things. Just as video games emerged from the context of computer research in the middle of the twentieth century, so did modern affect theory, in the figure of American psychologist Silvan Tomkins. My take on affect draws on Tomkins's decades-long work on a theory of the affects as well as on the ways his work fits into contemporary debates about affect theory. Tomkins's cybernetically influenced ideas about the coassembly of affect and cognition are often figured as counterpoints to theories of affect inspired by the work of Gilles Deleuze that are invested in a sharp move away from subjectivity and representation. One of my primary claims in this book is that by historicizing affect theory and thinking through how it and video games are the conjoined legacy of cybernetics, we might move both beyond the problematic impasses that were produced when computational metaphors ceased being metaphorical and the language of cybernetics, the computational sensorium, and everyday feelings transmitted across bodies and machines became meaningfully interwoven.[10] In doing so, we might enable textual and procedural approaches in game studies to speak more convincingly to each other about how games make complex meanings across history, bodies, hardware, and code. And affect theory might gain a historically and materially rich example of how cybernetic systems like video games came to underpin affect's claim to the body in the first place. In short, in this book I make a case for why media theory is not finished with representation and subjectivity.

We can understand video games—their historical and technological underpinnings and their contemporary ordinariness—as giving expression to our affective harnessing to computational processes beginning in the middle of the last century. In Williams's later work on "structure of feeling," he argues for an approach to cultural theory that leaves room for the possibility that media and art might give expression to emergent shared feelings that are not yet present in language, but in which we might sense the rhythms and emotional tones of new ways of being in the world.[11]

Williams's "structure of feeling" points toward what only later would come to be called affect theory. Contemporary affect theorists have taken up Williams's concept for the way in which, in its time, it served as a structure of feeling for an emerging dissatisfaction with structuralism's limitations. If video games can be interpreted as structures of feeling—as giving expression to emergent ways of being in the world—then affect theory, too, might be interpreted as giving shape and expression to the technosocial conditions under which we came to need a conceptualization of "emergent feelings."

Video Games

Video games are affective systems. When we open a video game program on a phone, computer, or gaming console, we are opening up a "form of relation" to the game's aesthetic and narrative properties, the computational operations of the software, the mechanical and material properties of the hardware on which we play the game, ideas of leisure and play, ideas of labor, our bodies, other players, and the whole host of fraught cultural meanings and implications that circulate around video games. From pressing buttons on a controller to navigating through a virtual world, feeling pleasurably immersed in a landscape or mission, to tapping and swiping brightly colored candy on a mobile phone to pass the time on the commute from home to work, video games give color, rhythm, shape, and sound—a texture and a tone—to time spent with everyday computational systems. Video games ask us to make choices, and they ask us to operate within the sets of constraints or rules that govern those choices. Video games ask us to understand, on a cognitive level, the underlying logic of their systems. They also engage and entangle us in a circuit of feeling between their computational systems and the broader systems with which they interface: ideology, narrative, aesthetics, and flesh. In this way, video games as pervasive and popular media are uniquely suited to giving expression to ways of being in the world and ways of feeling in the present that can tell us something about contemporary digitally mediated and distributed subjectivity.

I use a purposely broad definition of "video games" in this book—from the narratively complex and graphically rich products of the contemporary video game industry to comparatively simple mobile games, and from early "computer games" to experimental ludic works by artists. Many of the problems in game studies and in gaming culture that this book speaks to can be attributed in part to overly narrow and ahistorical debates about what a video game most essentially *is*. Video games, in their many shapes and sizes, are ubiquitous in our everyday media landscape. Their structures and aesthetics inform everything from cinema and contemporary art to pedagogical strategies and self-help manuals. Global video game sales vastly outpace film and recorded music sales combined. Video games are found everywhere in public, yet they are also very private, pleasurable and contested, extraordinary and banal, one thing and many things at once. They can be expensive or cheap, violent or sweet, flagrantly stupid or philosophically sophisticated. Video games offer ways of experiencing the time and space of contemporary life that are different from those provided by other screen-based media. In this book I present an argument for understanding video games not as an entirely new medium or as an autonomous art form but rather as part of the historically and technologically grounded, yet emergent and evasive, shifts in the everyday conditions of our computer-mediated world.

Following the lessons of Williams, I argue that video games must be understood as more than just ideological training grounds for capitalism. While it is important to trace the ways video games harness our bodies to the twenty-first-century rhythms of labor and leisure in particular ways, we must also account for the aspects of video games that cannot be reduced to the military–entertainment complex. This book's conversation between video games and affect theory is also a response to modes of analysis that instrumentalize video games as beacons of social change.[12] Proponents of "gamification" tend to overstate what video games can actually do in culture. Insisting on the social value of games above all else results, as Miguel Sicart argues, in shallow and narrow theories of what "play" is.[13] Similarly, claiming video games for sociality and productivity tends to render as "bad" other types of games and play that do not easily fit this model. As Adrienne Shaw argues, game scholars might apply the lessons of the antisocial turn in queer theory to "more fully embrace precisely that which makes game play so irredeemable within the norms of reproductive, modern capitalist culture."[14] Video games are ordinary. By "ordinary" I do not mean normative, though many video games participate in heteronormative discourses. Starting from ordinary means that we treat video games as part of culture, as media that we live through and with in various and complex ways. Like Shaw's call for queering game studies, this approach means holding a place for nuance in video game analyses, holding a place for the more interesting possibility that video games are neither merely repressive nor liberatory but rather "a whole way of life."

This book builds on approaches in media and affect studies that attempt to name and analyze how feelings move through and get expressed through particular objects. Identifying a video game as an affective system means resisting locating properties like texture, tone, and feelings in a purely subjective experience of reception or as the exclusive property of a text, and instead locating them in the slippery and intellectually fraught place in between. Game scholars mining this in-between space face particular challenges. The early history of game studies was dominated by a debate that pitted approaches that focused on how video games remediate the conventions of literature and film against approaches that highlighted the ludic properties of video games as central to their meanings.[15] This debate subsided as scholars found ways to address the unique properties and history of video games without denying the influences of other media. Game studies found "proceduralism," an approach that locates video games' unique expressive potential in the interplay between their computational systems and the systems they simulate through rules, narratives, and mechanics.[16] At the center of proceduralism is a formalist approach that ascribes the essential meaning of a video game to how effectively and creatively its designers harness code to simulate a systemic relationship between the world of the game and larger philosophical issues. Yet the player as a feeling and active subject goes missing from strict proceduralist accounts, or is there only as the thing that pushes the buttons to activate the code. Theorists of play have justly criticized proceduralist approaches on this count.[17] Yet play-centered approaches to video games have their own blind spots. They inherited the ludologists' overestimation of interactivity and the power of players to invest video games with whatever meanings they choose. Such approaches also tend to overlook the specificity of particular video games as cultural documents and artistic expressions. I do not rehearse the narratology versus ludology debate here. This debate, however, shaped the field by creating an artificial divide between the representational aspects of video games and the games' computational architecture. In this book I reposition the debate to consider how narrative structure—its presence or absence in games—was never really the contested issue; rather, it was the *image* and the relationship between representation and subjectivity that procedural and play approaches pushed to the background.

Game studies has a problem with representation. In the focus on interactivity and code, we have lost some critical tools for analyzing how video games matter as representations and how they are bound up with contemporary subjectivities. The field's move away from representation can be understood in part as a reaction to the ways representations of violence in video games have historically come under attack by politicians and "family values" watch groups. It is easier to say that video games are not a representational form than to wade into the thorny issues around "media effects." But this has left game studies lacking a compelling response to the moral panics. We have retreated into computation. This has produced necessary and important work, but it has also left game studies ill equipped to address issues like racism, homophobia, and misogyny in video games and gaming culture. The computation/representation divide in game studies validates and reinforces these problems. The implicit message of the field is, unless you can ground your analysis of race or gender in code or the interface, you had best not bring the subjects up.

While this book is a critique of this divide, it is also an intellectual history. I believe game studies is already starting to move beyond this divide, but by naming and historicizing it I also hope to offer the field some new routes through these problems. My approach to game analysis in this book is aligned with and builds on recent work in game studies that accounts for the complexity of meaning making in games beyond proceduralism and play, without regressing into these previous debates. Game scholars Clara Fernández-Vara and Christopher A. Paul, for example, make compelling cases for the importance of textual analysis to game studies. The cultural significance of games emerges from the context of play, Fernández-Vara argues, and also "results from the player interacting with the systems and representation of the game."[18] Similarly, Paul argues for analyzing all of the rhetorical aspects of games, not just the procedural rhetoric, to understand "how games persuade, create identifications, and circulate meanings."[19] What this book adds to these developments is an approach that moves across and productively integrates the text/context, core/surface, and computation/representation binaries on which so much video game scholarship is stubbornly premised. Affect theory, in that it mines that space between the expressible and the inexpressible, provides a way to read across code, images, and bodies without reducing video games to either their representational qualities or their digital and mechanical properties. Taking up affect in video games allows us to see the homologies among the actions of a player's body, the actions of a game's mechanics, and the actions of ideological signification to get at how video games and gamers, as particular cultural formations, are affectively charged. Thinking though affect and video games allows us to see how game mechanics are intimately tied to their representational practices and that game mechanics, code, and software are themselves also kinds of fictions that we tell ourselves.

Back to the woman on the train: *Candy Crush Saga,* with its colorful Candy Land–inspired world and constant in-game appeals to the player to make micro-purchases, is perhaps the most ubiquitous and banal example of the medium. It is partly for its ordinariness that I begin this book with it and then take it up again in chapter 3. It is also here because it is a game, like all the video games in this book, that I have played a lot. The games that I analyze and my theoretical framework are an "orientation," in Sara Ahmed's sense of the term.[20] That is, my choices of games and my understanding of how games make meaning cannot be meaningfully separated, because both are informed by my affective orientation toward certain games and away from others. This, of course, has its limitations, and I call it out here because this has been the unnamed starting place for a great deal of video game theory that, until quite recently, has ignored the limitations of its own surprisingly homogeneous orientation toward big-budget games. This has meant, as Jon Dovey and Helen Kennedy argue, that the field and the types of questions it asks have been bound up with a fairly narrow conception of what "video games" are, who plays them, and why.[21] As a result, the qualities of casual games, indie games, and art games that put pressure on these definitions have been obscured. Even more so, this lack of reflexivity in game studies contributes to the validation and reproduction of the raced and gendered hegemony of the tech industry and the perception of gaming culture as the same.

In this book I offer a feminist intervention into intellectual formations that have implicitly (and sometimes explicitly) rendered certain types of games, certain aspects of games, and certain types of critiques as "feminine" and therefore less rigorous and less valued. By focusing on both smaller games and affect, I am doubling down on the easily dismissed. One of my main contentions, discussed at length in chapter 1, is that the links between cybernetics and affect theory in the mid-twentieth century had explicit effects on the computational systems (especially games) and the social scientific theories that emerged from that moment. There are intellectual links between this history and the current field of game studies. Influenced by "new materialism" in media studies, game studies has decentered the subject and her relationship to images in order to privilege "thingness," proceduralism, and action. As evidenced by the Gamergate attacks on women game designers, players, and critics, such arguments in game studies actually give validation to those who reject any attempt to see video games as addressing particular subjects or as engaging in the politics of representation. "Gamers" often reject feminist critique on the grounds that games should be evaluated not as representations but rather as playful and apolitical computational systems. Using theories of affect that are committed to the subjectivizing and collectivizing force of media, I argue against the possibility of even making such a distinction.

By focusing on the affective qualities of games that exceed algorithms and aesthetics, I seek to expand our understanding of the ways video games and game studies can participate in feminist and queer interventions in digital media culture. What unites the various types of games in this book is an approach that reads them as affective systems and that takes into consideration how bodies, code, hardware, images, sounds, and sociohistorical contexts work together to give shape to feelings that exceed any one of these locations or modes of interpretation. Putting theories of affect in direct conversation with video games zeroes in on the contact zones between bodies, platforms, images, and code to redress the atomization of these approaches in new media studies.

Affect

Considering video games as structures of feeling necessarily puts pressure on some of the ways affect has been theorized. In this book I draw on a variety of theorizations of affect—not all necessarily in perfect agreement—in order to tease out what is useful about each for understanding video games as part of our contemporary sensorium. "Affect" is a notoriously slippery concept.[22] This is perhaps not surprising for a concept that seeks to encompass something as complex as the forces that inform our emotional states. This—forces that inform our emotional states—may be the most basic definition of affect shared by most approaches.

I use the term *affect* in this book to refer to the aspects of emotions, feelings, and bodily engagement that circulate through people and things but are often registered only at the interface—at the moment of transmission or contact—when affect gets called up into representation. We can attribute the affective turn in theory in part to dissatisfaction with the emphasis on the individual subject in psychoanalysis and the emphasis on language in structuralism. What affect names is a way of talking about the myriad ways everyday experience is felt but is not articulated or is inarticulable. But important to my discussion here is the way affect also speaks to the ways these felt but unexpressed feelings might be held in common and also might stem from actual, material, and nameable conditions.

Of considerable influence on this work are feminist, queer, and subaltern theories of affect that emphasize it as a social, subjectivizing, but also collectivizing force. The work of Eve Kosofsky Sedgwick, Lauren Berlant, Sara Ahmed, and Sianne Ngai is particularly useful for tracing the ways emotions have a social and political force that escapes conventional analyses of individual psyches and personal experiences. Affect in their work is that which both restricts and makes possible the notions of personal, collective, and emergent identities. For these theorists, affect is a deeply relational force that attaches itself to and is expressed through all kinds of cultural texts. Affect shapes the surface and very being of subjects and objects as they come into contact with each other. Affect is not inherently counterdiscursive or progressive; often, shared feelings tend to reinforce individual, regressive, and normative ideas. But affect also circulates and is subject to history and change. Affect has the possibility of forming counterpublics around the cultural expression of underrepresented feelings.

Central to my argument is an examination of the parallel thinking that informs gendering of game studies along a computation/representation divide and the gendering of affect theory along a neuroscience/culture divide. As such, my approach is a critique of Deleuzian versions of affect theory that suggest that it liberates theory from dealing with representation and liberates the sensing body from a grid of signification.[23] The claims made by such theorists suggest that the body in movement and the virtual capacity of affect model a new politics and a new ethics. Yet this potential, this becoming, is never actually here. Once present and comprehensible to us as "feeling," it has entered into the realm of signification and representation. It has been stilled, and, in its stillness, confined to the grid, it loses its radical potential. These theories of becoming are compelling for their hopefulness, yet we can rarely apply them to particular subjects or objects because in doing so we always seem to slip into the problem of language that these theorists wish to avoid. How do we grasp this virtual potential? How do we form alliances in movement? Despite the effort to use affect to describe a new type of relationality, it all seems to come down to the unique space-time of individual bodies (human or otherwise) making their way through the world—the uniqueness of bodies dancing to their own beats, their own processual rhythms, only occasionally coming into contact and harmony with other bodies.

I am more interested in how bodies come to feel similarly across and through objects and ideas, how these shared space-times—or rhythms—have shifted as the rhythms and sensual properties of our media have changed. I am interested in affect not as a virtual process running in the background or alongside bodies, but as the embodied capacity to feel—that which simultaneously opens us up to the world as relational beings and reminds us that our own sense of individuation and connection is always partial and extremely limited. Affect stills us, gives us shape, identifies, signifies—as *this* and not *that* or this *and* that, even if these terms are shifting—as toward this object and oriented in space-time even if, yes, the body is always moving. Rhythm requires both movement *and* stillness. At the interface we can see these rhythms. A woman killing time on a subway platform with *Candy Crush Saga*; commuters being alone and together while playing similar games on their phones. In such moments of being in relation through a type of signifying structure, we do not lose sight of affect; rather, this is the only possible way to make sense of it. The rest of the time it is too blurry and diffuse. At the interface we get fragments that tell us something about the larger picture that cannot all be grasped at once. A video game is such an interface for grasping a contemporary structure of feeling.

Some scholars of affect have turned to Tomkins to provide a counterpoint to Deleuzian versions of theory that emphasize affect as untethered from language and mediation.[24] Tomkins keeps us in the realm of feelings tied to individual bodies and their expression in language, their social formations and meanings, but he also allows for affective processes that exceed what any individual can know and express, tapping into feelings as they move across bodies, and their function in making certain relations possible and others not. Tomkins's theory of the affects is far from perfect, but his language is humane in its promiscuous blending of the biological and the social, in the *what is* and the *what might be* possible when we take feelings seriously. Perhaps most significant for the ideas in this book, Tomkins's theory of the affects, which emerged from the cybernetic moment that also birthed video games, holds embodiment, sociality, and technology together. Thus Tomkins provides both theoretical and historical insights into our experience of video games in the present. In taking up Tomkins to think about video games as affective systems, I am more interested in the historical specificity of his theory than in an application of his neurological framework. Tomkins's notion of affect as a "mattering mechanism" is particularly apt for exploring the larger significance of why we play games and why they matter, how we move through their virtual worlds and how games move us.

In this book I develop these ideas through chapters that consider the affective aspects of historical disorientation, everyday entanglement with machines, rhythms of work and play, and the less ordinary potential of games to break us out of these rhythms and contort our bodies in the arrhythmic perturbations of failure. In chapter 1, I establish the historical foundation for bringing affect theory into conversation with video games and vice versa. I use the contemporary adventure game *Kentucky Route Zero* and Tomkins's theory of the affects as historiographical lessons through which to consider the affects of orientation and disorientation in video games and in history. Through discussion of Laine Nooney's call for "spelunking" as a way of doing feminist video game history, I grope around in the darkness of *Kentucky Route Zero*'s caves to analyze how the game, like the legacies of cybernetics with which it is concerned, productively confuses distinctions between self and other, inside and outside, surface and depth.[25] This, I argue, provides a model for affectively reorienting us toward video games, computer history, and their limitations, particularly the bodies, objects, and stories rendered invisible by them. Chapter 2 takes up how surface/depth distinctions in affect theory map onto representation/computation distinctions in game studies. What does it mean when a game touches us and what does it mean when we touch a game? Focusing on the screen, specifically the touchscreen, and its images and materiality in video games permits an engagement with the feeling of our everyday entanglement with digital devices. Thinking through affect at the ludic interface, we see how our bodies, images, and code are meaningfully entangled. Representation has always been an embodied experience, and representation matters, *is* matter. Chapters 3 and 4 consider the ordinary and extraordinary potential of video games to put us into affective states that are useful for dealing with contemporary digitally mediated life. In chapter 3, I look at casual games played on mobile devices and the rhythms of work and play and their gendered dimensions. Casual games are ludic interludes that remediate the experiences of flow and interruption in the workday and tap into a longing to feel differently about work. Chapter 4 turns to arrhythmia and failure. How might video games meaningfully pull our bodies out of the everyday rhythms of digital labor and permit us to reflect on the centrality of failure under capitalism?

Two of the primary claims I make in this book are that bodies are not machines and that affect is not virtual. This is not a restating of liberal humanism in the face of posthumanist thought; it is more about the peculiar discursive erasures that occur when our machine metaphors collapse and become fact. We lose the nuance and potential of important differences. In the pages that follow, I try to recapture some of this nuance and some of these differences.

CHAPTER 1

Feeling History

CAVE CITY, KY—*Subterranean explorers announced today their discovery of a long-sought link between two of the world's major cave systems, which stretch for 150 miles underground. The discovery was made three months ago by a party of five men and a woman that spent 16 hours, often in water up to their necks, worming seven miles underground to locate the long-suspected connection between the Mammoth Cave and the Flint Ridge Cave systems. . . . Mrs. Patricia Crowther of Arlington, Mass., a 29-year-old computer programmer who is the mother of two daughters, told in an interview of being soaked to the skin, caked with mud "like chocolate frosting" and almost exhausted, when the party inched through Hanson's Lost River and into the Mammoth Cave complex last Sept. 9. . . . Mrs. Crowther, described by her colleagues as a "gung ho caver," was credited by them with leading the way to the discovery. During a 21-hour trip on August 30 the 115-pound Mrs. Crowther was able to squeeze through a narrow canyon on which was found scrawled the name "Pete H." together with an arrow pointing toward Mammoth Cave.*

—New York Times, December 2, 1972

THOSE FAMILIAR WITH VIDEO GAME HISTORY might recognize the name Crowther in the context of caving in this chapter's epigraph. William Crowther, also a caver, computer programmer, and Patricia's ex-husband, would write the text-based computer game *Colossal Cave Adventure* just a few years after this newspaper article was published.[1] The game would become one of the most iconic, copied, and beloved games in video game history, inaugurating the "adventure game" genre. Crowther has said that he made the game as a way of staying connected to his daughters after his divorce from Patricia.[2] He wanted to share with them his dual passions for computer programming and spelunking, neither of which his young children could participate in directly. *Colossal Cave Adventure* takes place in a massive cave system designed by Crowther to be nearly identical to Kentucky's Mammoth Cave system (though embellished with magic, treasure, and dragons). The game created an affective interface between cave systems and computer systems, between work and leisure, and between a father and his daughters.

How can a video game affectively reorient us toward history? What potential does asking this question have for feminist game studies? How might we account for the ways affect circulates through the corporeal and discursive bodies in this small news item from 1972 that retroactively pulls an iconic video game into its emotional field? Patricia Crowther's exhaustion and elation as she pulls her slight frame through the narrow gap; William Crowther's love for his daughters while he writes a game program late into the night; a newspaper article that ascribes the identities of computer programmer, mother, and gung-ho caver to the same body; Patricia's description of the cave's mud as "chocolate frosting," collapsing the danger and daring of cave exploration into the motherly space of the kitchen. Relatedly, how might game studies account for the embodied and discursive ways affect circulates through video game history?

Women's creative and consumptive practices of digital technologies have not been legible in gaming culture's attachments to hackers and hardware. Similarly, approaches to video game analysis that privilege mechanics and code (as the "proper" domain of programming) over images, characters, and story (rendered as secondary to the action) have the effect of foreclosing other types of analyses, and other types of players, games, and reasons for playing, that might differently attune us to how games make meaning across bodies and code. This story about *Colossal Cave Adventure* and a computer being creatively reimagined as a portal to another realm of mud, darkness, humor, and play, where love might be communicated through code, does not seem to fit most accounts of what early video game culture was like. It seems divorced from the parade of platform wars, blockbusters, and technological innovation that constitutes most popular and scholarly accounts of video game history. Where does Patricia Crowther—a programmer, mother, and caver—fit into this history? Although the digital computer takes its name from the women who worked as "computers" prior to the 1940s, women's bodies and histories are scarce in postwar histories of computer technology. This historical invisibility is compounded by the postwar misrepresentation of women's labor in the computer industry that renders the contributions of women as less creative, complex, and meaningful than those of their male counterparts.[3] Video game histories and cultures inherited these practices and occlusions and made them worse in many ways.

In response to this problem, Laine Nooney has called for "spelunking" as a way of doing feminist video game history.[4] Nooney uses the word spelunking to signal a slight shift from media archaeological approaches. An archaeological approach is necessary, Nooney argues, to undo teleological timelines of video game history, but something else is needed to avoid simply "adding women" or "envisioning histories and theories without corporeal or discursive bodies, histories or theories lost in their own love for the mechanism's indifference to the body." Spelunking, for Nooney, is not about digging to reveal an obscured history but rather about groping in the dark to sense the embodied and structural limits of what media histories and analyses can reveal. It "is a phenomenologically imprecise encounter—I can only see so much at any one time. The shape I hollow out here relies on non-continuity and the inability to apprehend the historical field in its wholeness."[5] Feminist approaches to video game histories and theories should reject those that present mythologies of completeness and teleological timelines. Staying in the dark, looking at fragments, feeling our way around the neglected objects of video game culture and overlooked modes of play might produce feminist gaming histories and cultures that are more expansive than what Lara Croft's body signifies. Patricia Crowther did not design a game (that we know of), but using the affordances of her slight frame, she maneuvered her wet, mud-caked body through the dark, narrow underground passages through which her companions could not fit. To spelunk is to explore a potentially vast space that can be apprehended only a small section at a time. To spelunk is to risk disorientation in space and time. In a cave, our access is limited by the contours and qualities of the space, the contours and qualities of our bodies, and the affordances of the technologies at hand. Also, through attention to the particular affective and technological affordances of video games, we might gain new approaches to history more generally.

In this chapter I go spelunking through two different, but related, caves. First, I explore the caves of the video game *Kentucky Route Zero* (Cardboard Computer, 2013–), a contemporary homage to *Colossal Cave Adventure*. I argue that *Kentucky Route Zero* can be read as a feminist historiographical object in that it is a game that reframes video game and

computer history as a series of choices made within the complex interme-
diation of technology and subjectivity in the latter half of the twentieth
century. The game uses the self-reflexive humor of the adventure game
genre—humor that is premised on the limitations of programming—to
enfold time and space, fact and fiction, and bodies and technology. These
enfoldings and disorientations affectively reorient our assumptions about
these categories and their relationships and open up spaces for newly
imagining women's contributions and relationships to video games. As
such, it is a game that recognizes the imbrication of corporeal and discursive bodies—actual bodies and represented bodies—in how history is
told and felt. *Kentucky Route Zero*'s narrative and mechanics foreground
the act of groping around in the dark, and the potentials and pitfalls of this
process, as a historiographical method that is less about revealing some
obscured truth and more about "sensitizing us to more complex historical
textures."[6]

Second, I explore the links between affect theory and game studies
through what Eve Kosofsky Sedgwick and Adam Frank call the "cybernetic
fold" of the 1940s–60s.[7] This fold—a cave of sorts—consists of a reconsideration of American psychologist Silvan Tomkins's cybernetics-inspired
theory of the affects. Sedgwick and Frank figure the fold and Tomkins's
work as historical and conceptual spaces where corporeal and discursive
bodies overlap in ways that bypass the binaries of structuralism. Revisiting
this argument and Tomkins's theory of the affects, I explore the historical
confluence of video games and affect theory in the mid-twentieth century
as an important historical, social, and technological assemblage that in-
evitably shaped video games, affect theory, human–computer interaction,
and the discourses around them. I draw out the connections between affect theory and video games and their simultaneous emergence from attempts to communicate feelings between humans and computers in the
context of a shifting and unsettling technological landscape. In doing
so, I demonstrate that strains of affect theory that implicitly borrow the
machinic language of cybernetics without historicizing these metaphors
cleave affect from language and subjectivity to reinforce the same impasses
they seek to redress—specifically, digital/analog, human/machine, and
reason/emotion. In Tomkins's theory of the affects, in contrast, we find
a model for moving across and through these impasses without ignoring
their structural complexity—not collapsing the human into the machine
but also not suggesting these things can be neatly separated or even realis-

tically disentangled in any approach that takes bodies and media seriously. In using affect theory as an approach, feminist game studies must carefully historicize the particular language of affect it uses.

This chapter addresses both how game studies might better account for its "own love for the mechanism's indifference to the body" and how through video games contemporary theories of affect might gain a more robust understanding of how the computational imaginary of the twentieth and twenty-first centuries profoundly underpins affect theory's particular claim to the body in the first place. Before descending into these caves, however, I turn to feeling disoriented as a method for video games and history.

Groping around in the Dark

Nooney calls spelunking a "phenomenologically imprecise encounter." It takes into account the limitations of historical interpretation and recognizes that, given the subjectivity at the center of every history, imprecision must be foregrounded and analyzed rather than disavowed. This approach makes claims for the value of fragments over any sense of a whole picture. Nooney argues that such an approach is necessary for feminist game studies if we are to go beyond "widening" history to actually account for the ways game studies partially creates the conditions for the marginality of particular bodies in the first place. She applies this approach to the career of Roberta Williams. Inspired by adventure games like *Colossal Cave Adventure*, Williams cofounded the software company Online Systems (later Sierra Online) with her husband, Ken, and over the 1980s and 1990s created some of the most popular and influential computer games of that era. Using some fragments from Williams's career—magazine articles, memories, fan letters—Nooney demonstrates how uncomfortably Williams's body, creative practice, and games fit into the narrow contemporary definition of what and who counts in game history. These fragments do not unearth and restore a "missing" aspect of video game history; rather, they reveal the way "our sense that videogame history is 'all about the boys' is the consequence of a certain mode of historical writing, preservation, memory, and *temporally specific affective attachments,* all of which produce the way we tell the history of videogames."[8] Williams was not a programmer or part of the California tech scene; she saw her game design practice as an extension of her interest in literature and storytelling. "Insofar as videogame history struggles to represent itself as much more than a chronology of consoles, games, and programmers," Nooney argues, "the field fails to critically inquire into the ways *gender is an infrastructure* that profoundly affects who has access to what kinds of historical possibilities at a specific moment in time and space."[9]

Analyzing how a video game feels and how it can put us into a particular affective relation to history is a "phenomenologically imprecise" method. It opens game studies up to other affective attachments, histories, and ways of being in relation to video games. Phenomenological imprecision is, in some ways, an apt description of the myriad approaches and debates that constitute "the affective turn" in theory. Imprecision is also a common critique of affect theory—that its language of intensity, becoming, and in-betweenness and its emphasis on the unrepresentable give it a maddening incoherence, or shade too easily into purely subjective responses to the world. But imprecision also names one of the fundamental stakes of affect theory's intervention: that the legacies of structuralism and poststructuralism created theoretical impasses built around the imprecision of symbolic systems in wholly accounting for human experience.

In *The Forms of the Affects*, Eugenie Brinkema somewhat hyperbolically claims that the affective turn in film and media studies has abandoned the field's major strength—attention to form—and has ceased interpretive work altogether in favor of tracing elusive and subjective "feelings" in the spectator that are somehow communicated through and across bodies and media objects. She writes: "Film-theoretical accounts of the nexus of terms 'emotion,' 'feeling,' and 'affect' have not strayed far from the dominant Western philosophical models for thinking about interior states or the passionate movement of subjects."[10] Brinkema rejects this "expressivity" approach and instead argues for a "radical formalism." "[In] an attempt to dethrone the subject and the spectator [i.e., bodies]," Brinkema writes, "affect will be regarded as a fold, which is another way of saying that affects will be read for as forms."[11] She uses the concept of the fold to make a case for "reading for affect" in film and other visual media. Affect is a formal quality rather than a subjective emission or expression from a body. The "fold" as theorized by Gilles Deleuze is a material phenomenon that by its very structure calls into question what is inside and what is outside.[12] In a folded form, what appears as inside is just the outside for another inside.

The fold is useful to the degree that both affect theory and game studies have struggled with notions of "expressivity." Are emotions generated somewhere deep within and expressed through gesture, facial contortions, and language as secondary systems, or vice versa? Are the *real* meanings of video games found in their invisible layers of programming, rather than in the expression of that programming in images and sounds? The answer the fold might seem to provide is the circumvention of the inside/outside binary to suggest a more nuanced structure of expressivity. But instead, the fold in Deleuzian strands of affect theory has been used to circumvent expressivity and bodies altogether.

I share with Brinkema an interest in reading for affect. In fact, I am convinced it is the only method we have for apprehending something so fugitive. I read video games not as containers of and for affects that float around between bodies and things but rather as media that have specific affective dimensions, legible in their images, algorithms, temporalities, and narratives, that can be interpreted and analyzed. I concur with Brinkema when she writes:

> Affect is the right and productive site for radically redefining what reading for form might look like in the theoretical humanities today. First and foremost, this approach requires beginning with the premise that affective force works over form, that forms are auto-affectively charged, and that affects take shape in the details of specific visual forms and temporal structures.[13]

I want to put pressure, however, on how Brinkema gets to this point—specifically, how her use of the fold abuts but disavows the legacy of systems thinking, the cybernetic fold, in contemporary theory. What I propose in this chapter is that this is a partly historical problem rather than a wholly theoretical one. The kind of structuralist theory that emerged out of the mid-twentieth century implicitly used the threat of computers to draw rigid boundaries between affect and cognition, humans and machines, the virtual and the actual, the digital and the analog. Yet even critiques of this, like those that emerge through Deleuzian notions of affect that position it as that which exceeds theory's linguistic traps, use the same cybernetic language and systems thinking from which these perceived traps emerged.

Released a few years after Sedgwick and Frank's analysis situating Tomkins's work in the "cybernetic fold," Brian Massumi's *Parables for the Virtual* builds on Deleuze and proposes that affect is a virtual force that exists outside subjectivity and, hence, language. Affect, for Massumi, is a quality of the "processual" body that is presubjective. To get to this understanding of affect, however, Massumi oddly removes affect from the body. He describes it as an "incorporeal dimension of the body"—something related to the body but not entirely of it.[14] Affect is part of a "processual indeterminacy" that precedes consciousness, subjectivity, and language, and as such it is "primary in relation to social determination," where positionality related to gender, race, and sexual orientation is secondary and derivative of this virtual realm.[15] Here, Massumi uses the language of cybernetics—"feedback" and "processing"—to cleave affect from subjectivity, feelings from bodies, expressivity from representation. In her critique of Deleuzian strands of affect theory—primarily Massumi's work—Ruth Leys argues that this type of affect theory reinforces the same mind/body dualism in philosophy and critical theory that it seeks to call into question.[16] One of the primary stakes of the argument I develop in this book is to put pressure on a similar dualism in game studies and digital media studies more broadly. As discussed in the next chapter, this is the dualism that pits computation against representation. In this formulation, computation or code is the presymbolic, the sensation, and the unconscious, and representation is the symbolic, the cognitive, and the interpretive.

Massumi argues that twentieth-century theory's obsession with language created a discursive subject rather than a sensing subject. For Massumi, affect—a force that escapes language and other forms of representation—reattunes us to the corporeal body and its capacities, particularly motion and change, which cannot be called up into language. The body in movement and the movement of affect across bodies and things are forces, he argues, that cannot be accounted for through subjects fixed by the grids of symbolic systems. In this way, Massumi seems to be calling for something quite different from what Nooney suggests. Whereas Nooney is interested in how certain bodies are figured differently into video game history than others, depending on how legible or not their creative and play practices are to the existing historical methods, for Massumi "gender as infrastructure" is just another grid that foregrounds the discursive body over the sensing body. Massumi's premise, that the sensing body can be meaningfully accounted for by theory separately from the discursive body (the body of signification), is one of the fundamental problems that many feminists have with affect theory. The sensing body and the discursive body cannot be separated if one still cares about what it feels like to be made a legible subject through forms of representation. For feminist and queer theorists, the sensing body is a body that is made perceptible only through its relations to other bodies, other things, and other possible worlds. We are made and remade through our affective attachments to representations. This is not about fixing bodies into grids of signification and calling it a day but rather about thinking through how affect holds discursive and corporeal bodies together. We can analyze form, read for affect, and hold on to the body at the same time. Video games require this.

Affect is what puts bodies and/as forms into relation to each other in time and space. As Sara Ahmed argues, affect is a kind of orientation.[17] The need for phenomenological imprecision to account for gender as an infrastructure in video game history is related to Ahmed's queer phenomenology, in which she traces the ways different bodies are differently oriented (and disoriented) in relation to the world. She writes:

> What does it mean to be oriented? How is it that we come to find our way in a world that acquires new shapes, depending on which way we turn? If we know where we are, when we turn this way or that, then we are oriented. We have our bearings. We know what to do to get to this place or to that. To be oriented is also to be oriented toward certain objects, those that help us find our way. These are the objects we recognize, such that when we face them, we know which way we are facing. They gather on the ground and also create a ground on which we can gather. Yet objects gather quite differently, creating different grounds. What difference does it make what we are oriented toward?[18]

Playing with the language of "sexual orientation," Ahmed's words draw out the spatial and temporal specificity of bodies in relation to the world and how, in turn, this specific embodiment is given shape and dimension by proximity and distance, access and occlusion, facing and turning away. Ahmed's sensing body is not fixed by a grid, as in Massumi's critique, but its boundaries and movements—its abilities to feel and act—are at least partially determined and constrained by the markers of race, gender, class, religion, and sexuality. Here, the corporeal body can never be meaningfully separated from the discursive, symbolic, or representational body.

My particular orientation toward video games is one of fits and starts, hazy memories filtered through gendered and classed access to computers and home gaming systems and colored by generational and geographical specificity. Around 1981, when I was five years old, I remember occasionally being allowed to play games on the Atari system in my brother's bedroom. Since I remember little else about being five years old, clearly the games' sounds and images made an impression on me. Mostly, though, I remember watching my brother play. A bit later, in 1982 or 1983, playing *Pac-Man* on the Apple II at my uncle's house was also my first experience of using a personal computer. The Atari left when my brother left for college. It was not until 1986 that another gaming system would enter my house. For one brief and intense summer, I had access to a Nintendo Entertainment System (NES). A man who was attempting to woo my recently divorced mother (and, thus, also trying to win my favor) lent me his NES while he was traveling. I had only two games—*Super Mario Bros.* and *Duck Hunt*—but that was all I needed to become completely enchanted. Really, of course, it was the summer of *Super Mario Bros.*, but *Duck Hunt* had a certain function as a "break" from Mario. Even today, when I hear the *Super Mario Bros.* music, I can tap into the feeling of being ten years old and the way the game entrained my body, the tension in my hands and the contortions of my body in front of the TV as I mastered each level. That was it, though. My mother's friend returned and took his NES back. I did not play video games after that except occasionally at friends' homes or at birthday parties in arcades. I became a teenage girl, it was the 1990s, and, in my Southern California middle-class suburban milieu, this meant I did not play video games, except rarely and only ironically. Despite my lack of regular playing, I have always been freakishly good at *Mortal Kombat*. Once, in my early thirties, while on a road trip to Boston, I watched in awe while my similarly aged friend put a quarter in a rest-stop *Galaga* cabinet and played an epically long game on that one quarter. She probably had not played since she was eight.

Fragments. Turning toward and turning away. I am not a "gamer" in the sense of the gendered and other cultural baggage that that label carries, but video games—their presence and absence, proximity and distance—have mediated most of my life.[19] I returned to video games as an adult only out of intellectual curiosity while I was writing a dissertation on early experiments in human–computer interaction. As I discuss in chapter 3, my intellectual curiosity turned into using video games for small breaks and rewards during the lonely process of writing. Once I finished graduate school, I turned my scholarly attention to video games, bought various gaming systems, and began a project of what felt like "catching up." As I began teaching games and talking to more people about their gaming orientations, however, I realized that my path of a childhood interest in video games, a gendered lull in adolescence, and a return to gaming in adulthood is rather ordinary. Although I sometimes still feel like a fraud writing and teaching about games with this spotty history, with experience I have come to attribute this feeling more to the general "imposter syndrome" that plagues academics and less to something uniquely lacking in me.

This is my gaming orientation, and yours may or may not share some qualities with mine; the point about orientations is to understand their affordances and limitations. Ahmed's description of orientation sounds a lot like the experience of playing an adventure game. To play we must be conscious of where we are on the map and how to orient ourselves in relation to useful objects while also making decisions about turning away from other paths and other objects. For example, *Colossal Cave Adventure* begins with "You are standing at the end of a road before a small brick building. Around you is a forest. A small stream flows out of the building and down a gully." The first adventure games were text-based programs written for computers with command-line interfaces. In such a game, the player enters one- or two-word textual commands to advance, such as "east" or "enter building." The words orient the player in a space but provide little information about what to do or where to go in that space. The blinking cursor seems to wait mockingly for all the ways the impasses between human language and computer language might lead the player astray. Part of the puzzle is figuring out a text command that the program will recognize. Another part of the puzzle is recognizing the objects one might pick up along the way even if one is not sure how or if they will help. "To be oriented is also to be oriented toward certain objects, those that help us find our way. These are the objects we recognize, such that when we face them, we know which way we are facing."

Applying Ahmed's queer orientations to gaming culture, Adrienne Shaw argues that game studies needs queer theory to address the often homogeneous and unquestioned orientation of "what counts as play, who gets to play, and the assumed goals of play."[20] Following Ahmed and Shaw, then, I am making a case for feeling *dis*oriented—getting a bit lost and being unsure about one's position and what it permits—as a strategy for bringing into relief what we can see and say from a particular location and also for reorienting game studies in ways that might permit other meanings, other games, and other histories to emerge. Spelunking can be disorienting. Deep in the darkness our bodies can become unmoored in time and space. Through spelunking, groping around in the dark and feeling disoriented—the way this attunes our senses differently and leaves us open to the comic pratfalls and dangerous pitfalls of not seeing clearly—a model for apprehending the conjoined histories of video games and modern affect theory emerges. I argue for feeling disoriented in relation to history, and for the productive confusion between present and past, self and other, and inside and outside that both video games and affect theory might perform. In disorientation we might reorient ourselves as critics and scholars toward some of the binary impasses in contemporary digital and posthumanist theories. It is through such disorientation that we might also reorient gender in relation to video game histories, through a reconsideration of how the line demarcating computation from representation is also a line that demarcates boundaries around what and who count as video games' proper subjects and objects, historically and in the present.

Kentucky Route Zero

Kentucky Route Zero provides an intelligent and humorous model for feeling disoriented in history.[21] The game, set sometime in the 1980s (although the exact historical period is difficult to place), tells the story of a group of down-and-out characters in rural Kentucky who descend into a strange subterranean world—a large cave system—in order to find the address 5 Dogwood Drive. The game is set at first along a lonely strip of Kentucky highway that features a gas station, a farmhouse, a bait shop, and a mine; later, it moves to various odd locations along the mysterious underground Route Zero—a Bureau of Reclaimed Spaces, a Museum of Dwellings, a valley "bigger than a basketball court, but smaller than a financial district." The game foregrounds the experience of disorientation and, from this affective state, remaps video game history through the fragmented and distributed perspectives of several women. At first, the game appears to be about an aging truck driver named Conway, but as the player advances through the game, Conway serves as merely an entry point to a story about an artist and computer programmer named Lula Chamberlain. Chamberlain's story is presented in temporally disconnected fragments, mirroring Conway's (and the player's) disorientation in the game's world. Further, in order to traverse the game's confusing caves, Conway (and the player) depends on the technological ingenuity of his young companion, Shannon. A central concern of *Kentucky Route Zero* is the way our access to the past is always only partial, and uniquely so, depending on the limitations of the systems of memory and the bodies through which we store and access history, including the game itself. *Kentucky Route Zero* is also about how memory is distributed across people and objects, and how each person has particular affordances depending on how he or she is oriented in the world. These concerns are often made visually explicit in the game's compositions, in which the protagonists are dwarfed by vast and confusing spaces.

Kentucky Route Zero is explicitly indebted to the history of adventure games like those made by Crowther and Williams. Per the genre, a significant amount of gameplay involves reading description and dialogue boxes and making choices about how to respond. There are no specific gaming skills to be mastered except, perhaps, patience. *Kentucky Route Zero* is what contemporary gamers dismissively call a "walking simulator," which is meant to suggest that because it is not difficult to play, it does not qualify as a game at all. The "walking simulator" rhetoric of contemporary game culture is profoundly ahistorical. *Kentucky Route Zero* is, after all, a loving homage to one of the very earliest genres of digital games. As such, the game eschews realistic graphics and complicated mechanics in favor

"Random Access Self Storage." *Kentucky Route Zero* foregrounds the feeling of disorientation and archival concerns about historical memory in the information age.

of text, minimalist graphics, and point-and-click mechanics. It enacts an enfolding of geography, bodies, time, and technology to present an alternative and "underground" history of video games.

The cave in *Kentucky Route Zero* functions as a spatial and temporal metaphor for the game's creative reimagining of what lies "beneath" video games (their code and programming) or in their past (alternative and overlooked devices, figures, and approaches to virtual world making). At the beginning of Act II, one character asks another, "This might sound strange, but are we inside or outside right now?" The player is given a choice of responses:

CONWAY: Inside
CONWAY: Outside
CONWAY: Both

This confusion and the answer "Both" speak to the game's many enfoldings—in a cave, is one both inside and outside? The game's spatial confusion is doubled by temporal dislocation. The *when* of the game is difficult to pin down. Characters and timelines fold in on themselves. In Act III, a character is said to suffer from recurring feelings of "lateness." To the degree that the narrative invokes a history beyond the game, it is one that is layered with misdirection. Antique armchairs and old, broken television sets mingle with cell phones, e-mail, and references to the early history of the Internet. The dialogue, characters, locations, and images reference a dizzying array of real and fictional people: mathematicians, politicians, computer scientists, musicians, characters from literature, architects, and game designers. Most notable are the game's repeated references to major figures of cybernetics and computer science (Claude Shannon, Joseph Weizenbaum, William Chamberlain), early video games (Roberta Williams appears underground, working on the Xanadu system), and mid-twentieth-century literature (Samuel Beckett's *Waiting for Godot* and Gabriel García Márquez's *One Hundred Years of Solitude*).[22]

Part of the mystery of the game is trying to understand what exactly the relationship is between these actual historical figures and the story being told. The player is also trying to unravel the relationship between the world underground and the one that exists above, the internal and the external, the surface and what it obscures. The narrative is structured around a series of descents into the cave to locate the mysterious address, but there are also many diversions that bring the characters back to the surface. The themes of disorientation, structural limitations, and history are set up in the opening scene. The player's first task is to descend into the cavernous basement of the Equus Oils gas station to find the breaker box and restore the electricity. Looking for directions to his next delivery, Conway pulls his truck into a rural gas station and speaks to the elderly attendant, Joseph Wheattree. Sitting between the pumps in a Queen Anne armchair facing an old television set, Joseph tells him that he can only get there from "the Zero" and that Conway needs to find a woman named Márquez to direct him there. Márquez's address is on the computer inside the gas station, but the electricity is out; hence the descent into the basement. Conway's descent is represented from the impossible perspective of a cross section of the gas station's architecture, which reveals as he moves below ground that the building is shaped like an enormous submerged horse, of which only the head remains visible aboveground as the gas station's neon sign.

Before locating the breaker box in the basement, Conway comes across three people—Emily, Bob, and Ben—who are discussing the rules of a tabletop game that involves maps and a five-sided die. The moment serves several purposes. It is a nod to the immersive fantasy tabletop games like Dungeons and Dragons that have had significant influence on contemporary video game culture, especially for adventure games. The

Conway's descent into the basement of Equus Oils foreshadows the player's later descent into the cave and *Kentucky Route Zero*'s concern with what lies beneath video games and history.

five-sided die echoes 5 Dogwood Drive as well as the game's five-act structure, establishing the idea that the geographical, theatrical, and ludic structures of the game's story are linked. The tabletop game players do not hear
Conway when he asks them where the breaker box is, and once he restores
power and returns to the table, they have disappeared. In this way these
players serve as a spectral presence that self-reflexively gestures toward
the *Kentucky Route Zero* player, and also toward the idea that this game
is about the history of various games and various types of players. And
this all occurs within a strange horse-shaped structure that draws the player's attention to her orientation and the ways her movement and vision
are limited by both material and immaterial structures. Conway's descent
into the basement in the first scene foreshadows the other descents the
player will make into the game's fictional and historical caves. The game
first appears to be about Conway, in that he is the character the player can
first move and for whom she can make dialogue choices. But as the game
progresses, it dispels any notion of a stable perspective and distributes interactivity across multiple bodies, human and nonhuman.

Kentucky Route Zero's story hinges on the perspectives and technological ingenuity of the various women encountered along the way. At the end
of Act I, a moment of media archaeology opens the portal to the subterranean Route Zero, along which the characters discover an alternative history and geography of virtual worlds. The portal is created when a young
woman, Shannon (the name is a reference to the mathematician Claude
Shannon, whose digital circuit theory inaugurated the "information age"),
repairs a broken television set by replacing the cathode-ray tube with one
she scavenges from an old computer monitor. At another key moment,
Shannon repairs a broken radio, and that act of repair is what once again allows her and her companions to enter the Zero. It is Shannon's cleverness
that allows the player to move through the disorienting space and time
of the game. These acts of repair, and their significance in the narrative,
foreground the game's interest in the enfolded histories of mid-twentieth-
century technologies, the history of video games (and the way they were a
bridge between the television and the computer), and a reframing of gender in relation to the "hacker" persona and the conventionally masculine
histories of digital media. Interpreting *Kentucky Route Zero* as a historiographical lesson offers insight into how we might reframe game studies
in ways that permit more diverse feminist engagements with the medium

and how, in disorientation, we might be reoriented toward the way certain bodies and histories become visible in the dark.

What kind of an affective and historical (dis)orientation does an adventure game provide in the present? What *Kentucky Route Zero* offers is an imaginative reframing of video game history, one that more interestingly accounts for different types of origin stories and orientations within the entanglement of the natural and the mechanical that emerged from the mid-twentieth-century discourses of cybernetics, artificial intelligence, and structuralism that it revisits. To work this out, we need to move from the caves of Kentucky, actual and imaginary, to the "cybernetic fold" of the mid-twentieth century.

The Cybernetic Fold

Kentucky Route Zero's blending of computer history, video game history, and literary history positions these as historically imbricated sites of creativity and invention in the middle of the twentieth century. The game is distinctly interested in what Sedgwick and Frank call the "cybernetic fold"—the period roughly between 1940 and 1960 when cybernetics and systems theory were applied to everything from urban disorder and global politics to art, from the relationship between animals and their environments to the workings of the human mind.[23] Sedgwick and Frank make two related points about the cybernetic fold: first, that systems theory and structuralism are shared intellectual projects that were shaped by this moment, and second, that the shape that theory took in the latter half of the twentieth century was not an inevitable outcome of cybernetics but rather a particular response to the way computers and systems thinking called into question human autonomy and exceptionalism. The enfolded discourses of cybernetics, affect, and structuralism provide a context for the reframing of the emergence of video games during this same period and for contemporary analyses of the medium. Additionally, the cybernetic fold has lessons for how game studies and affect theory can productively speak to each other in the present.

The cybernetic fold, in Sedgwick and Frank's formulation, is the complicated historical moment when biological systems were overlaid with computational systems and vice versa. This layering of computational, ideological, and biological systems, they argue, shaped discourse in the twentieth century. The paths of thought and experience between the computer as a complex system, the body as a complex system, and consciousness as a complex system became entangled in ways that exceeded the power of metaphor.[24] This is what N. Katherine Hayles describes as the "Computational Universe."[25] According to Hayles, the legacy of this mid-twentieth-century moment is that *we now live in and as digital systems*—in the recursive loop created by the cybernetic fold. Similarly, Homay King argues that the computational metaphor and the "Californian ideology" it produced among the first wave of Silicon Valley entrepreneurs continued to cleave the body from intelligence.[26] King writes: "Digital media universes seemed to promise an alternate place of refuge from the weight and restrictions of Earth-bound existence. It was a virtual refuge, which would likewise require great feats of technical engineering."[27] As Fred Turner argues, the "computational metaphor" at some point ceased to be a metaphor as the power of computers, the ideological assumptions underpinning this power, and the institutional power of technology collapsed into each other.[28]

It is within the historical and theoretical context of the cybernetic fold that Sedgwick and Frank situate and revisit Silvan Tomkins's theory of the affects. Tomkins, who trained as a philosopher before turning to psychology, was interested in the nature of human will and was unconvinced that the rise of behaviorism adequately addressed this complex topic. Like Margaret Mead, Gregory Bateson, and many other intellectuals of this period, he was influenced by the proceedings of the Macy Conferences, a series of meetings held in New York City and sponsored by the Macy Foundation that began in the 1940s. At these conferences, researchers from a wide array of disciplines gathered to discuss the cybernetic and system theories of Norbert Wiener, Claude Shannon, and John von Neumann and how they might inform approaches in anthropology, economics, and psychology.[29]

Through cybernetics, Tomkins sought to describe the significant role that the affects play alongside cognition within a larger system of communication involving the skin, musculoskeletal, glandular, and neurological systems and social contexts. The "affects," which he grouped and listed as interest–excitement, enjoyment–joy, surprise–startle, distress–anguish, anger–rage, and fear–terror, are what Tomkins called "mattering mechanisms" that imbue sensory and environmental information with a tone or emotional texture that conveys to the cognitive system how to sort and

process that information. Tomkins's compelling contribution is the rec-
ognition that affect and cognition are mutually interdependent systems.
He writes:

> Because of the high degree of interpenetration and interconnect-
> edness of each part with every other part and with the whole, the
> distinction we have drawn between the cognitive half and the
> motivational half must be considered to be a fragile distinction. . . .
> Cognitions coassembled with affects become hot and urgent.
> Affects coassembled with cognitions become better informed
> and smarter.[30]

The affects for Tomkins are related to biological processes and needs but
are also deeply social. Our affective "scripts"—the feelings that arise for us
around particular scenarios, how we experience those feelings, and how
we communicate them—are not purely biological givens; rather, they are
part of an ongoing process of communication and interconnectedness
with our environment. The affects, in Tomkins's model, communicate
simultaneously inwardly and outwardly, both intra- and interpersonally. Affect, for Tomkins, is bound up with orientation. "Part of the power of the
human organism and its adaptability," he writes, "lies in the fact that in addition to innate neurological programs the human being has the capacity
to lay down new programs of great complexity on the basis of risk taking,
error and achievement—programs designed to deal with contingencies
not necessarily universally valid but valid for his individual life."[31]

Tomkins intended his theory as a challenge to the primacy of the
"drives" in psychoanalytic theory and also to the clear-cut distinction
between stimulus and outcome in behavioral models of psychology. In
cybernetics and computer science, Tomkins found a language of systems that he used to describe his theories about the way cognition and
the affects interact. Cognition and affect, for Tomkins, are interfacing
systems—a coassemblage. He writes, "Reason without affect would be
impotent, affect without reason would be blind."[32] In that he conceived of
affect as not purely biological, Tomkins took great interest in the emerging fields of computer science and extended his work to considerations of
how affect might inform the development of artificial intelligence (AI). In
1962, he presented a paper at a conference that he co-organized on "computer simulation of personality." In the paper, Tomkins proposed ways his

theory of the affects might help computer scientists develop AI. He argued
that in order for a computer to simulate human intelligence accurately, its
designers would need to create an affect system that could mimic human
sociality and learning across the life span.[33] In the present, his paper reads
as an argument for "machine learning," a model of AI that existed in the
early 1960s but did not really take hold until several decades later. In its
historical moment, however, the paper functioned more as a rebuke to the
simplistic accounts of human cognition that bolstered AI research. Human
intelligence, Tomkins argued, was not a data set that could be stored on
punch cards and communicated to a computer unless that computer also
had an interdependent program for learning complex human emotions.

In her book *Affect and Artificial Intelligence,* Elizabeth Wilson, like Sedgwick and Frank, returns to the cybernetic fold and to Tomkins to reani-
mate a particular lost intellectual moment when cybernetically inflected
theories of psychology might have gone in a different direction.[34] In the
context of the conference on computer simulation of personality, the
meeting of cybernetics and psychology was couched within discussion of
the possibility of using computers to research personality disorders. But,
as Wilson argues, Tomkins's paper and the conference proceedings had
wider significance in the field of psychology. After this moment, cognitivists dramatically disassembled the relationship between the body as a
biological system and the mind as a cognitive system. Cognitivists, led by
Ulric Neisser, defined cognition "as 'everything a human being might pos-
sibly do.' . . . Yet at the same time," Wilson argues, "the list of events that
might count as cognitive is circumscribed to '*sensation, perception, imagery, retention, recall, problem-solving* and *thinking.*' No feeling, no craving,
no growing, no hurting, no conflict or digression. What would count as
cognition would narrow dramatically in the years ahead."[35] The lost intellectual moment of the conference, according to Wilson, was when Neisser and others rejected the potential for continued research into "how
cognition and affect and information and motivation could interbreed"
and be operationalized computationally. "Neisser wanted his cognition
pure-bred (and purely human)."[36] The cognitivists' models and research
methods, however, were not purely human, in that they were deeply informed by the machine metaphors of cybernetics. Remarking on this relationship, Tomkins notes, "It appears that the regaining of consciousness is
less awkward for Behaviorists if it can first be demonstrated with steel and
punched cards that automata can think, can program, can pay attention

to input, can consult their memory bins, all in intelligent sequences—in short, that they can mimic the designers who intended they should do so."[37] This quotation about the unintended ideological effects and affects of computational systems applies just as well to current discussions about the ways the algorithms of social media reflect and broadcast the ideological biases of a small group of engineers in Silicon Valley.

Sedgwick and Frank argue that Tomkins's work also offered an alternative, less binary, path that structuralist and poststructuralist theory could have taken. They write, "The cybernetic fold is then the moment of systems theory—and also, in a directly related but not identical development, the structuralist moment."[38] Revisiting the cybernetic fold allows us to understand how computers—specifically, a promiscuous and loose understanding of how they worked—came to provide a compelling metaphor and ideological stance for the relationship between embodied consciousness and information. Theory in the latter half of the twentieth century, Sedgwick and Frank argue, might have looked different had other aspects of cybernetics and systems theory taken hold:

> The epithet "fold" seems applicable to the cybernetic moment partly because systems theory, precisely through its tropism toward the image of an undifferentiated but differentiable ecology, had as one of its greatest representational strengths an ability to discuss *how things differentiate*: how quantitative differences turn into qualitative ones, how digital and analog representations leapfrog or interleave with one another, what makes the unexpected faultlines between regions of the calculable and the incalculable that are destined to evolve into chaos theory.[39]

Cybernetics, then, was a way of seeing the world that emerged in a particular historical moment and that introduced a radical uncertainty into the very ideas of human intelligence and autonomy, only to be subsequently taken up by structuralist theory and computer science as a way of *processing away* that uncertainty to create symbolic order. What we get from this moment, Sedgwick and Frank argue, is an obsession in theory with systems of representation that are divorced from the body and an overdependence on language as fundamentally human and as the primary model for understanding representation. Theorists had to cleave apart the digital and the analog (and the related binaries of human/machine and information/representation) in order to shore up the shaky boundaries that seemed likely to erode sometime around 1950, when Alan Turing first asked, "Can machines think?"[40]

According to Sedgwick and Frank, poststructuralism inherited these boundaries, and even in trying to undo them, theorists did not interrogate the historically specific fear of human/machine miscegenation at their core. Sedgwick and Frank admit that theorists trained in the lessons of poststructuralist thought will reject Tomkins's work on many counts— biological essentialism, uncritical conception of the individual as a stable subject, a mechanistic model of subjectivity. Arguing against this, however, they point to how Tomkins's model of the affects resists these interpretations and moves beyond "fragile distinctions" such as nature/culture, mind/body, inside/outside, and self/other. But also, importantly, Sedgwick and Frank demonstrate how these interpretations are themselves symptomatic of the binaries and aporias poststructuralist theory inherited from the same historical moment. The coassemblage of affect and cognition and the blending of the biological with the mechanical, they argue, refuse a hard line between a notion of the mind as quantitative (digital) and the body as qualitative (analog). The layering and "leapfrogging" in Tomkins's model introduces queer possibilities for remapping human subjectivity along lines that are less clear and deterministic than those inherited from psychoanalysis and structuralism. The promise of Tomkins's proposition that "any affect may have any object," according to Sedgwick and Frank, is that it adds some productive opacity and polymorphousness to the relationships among consciousness, language, and subjectivation. Tomkins's model does not shy away from biological explanations for particular "innate" affect responses to stimuli, and it makes ample room for a dynamic and highly individuated understanding of how we are constantly undergoing complex and changing affective processes that are direct responses to our social contexts. For Sedgwick and Frank, the lost opportunity of the cybernetic fold is the failure to produce a more open-ended and less rigidly formalist twentieth-century theoretical discourse that could more fully account for the body as a social and biological system. Tomkins stayed in the zone "between regions of the calculable and the incalculable" and doubled down on Turing's question, asking in an ironic provocation, "Can machines feel?"

The legacy of the collapsed computational metaphor of the cybernetic fold is also borne out by some versions of contemporary affect theory, as discussed earlier. Understanding gender as infrastructure and affect as a kind of orientation contradicts some versions of affect theory that premise their interventions on either interpreting bodies without subjectivity or interpreting subjects without sensing bodies. For Sedgwick and Frank, the key to understanding the value of Tomkins's work in the present is its potential for conceptualizing affect and the idea of the sensing body in contemporary theory as a simultaneously historical, biological, technological, and social assemblage.

Video games, as cybernetic systems, need a theory of affect that can hold all of these together as well. Video games also emerged out of the cybernetic fold and were crucial in creating a new relationship between machines and bodies. Their emergence was premised on this moment of imbrication between cybernetics and subjectivity. They were part of a broader movement of using computer games during the Cold War to simulate complex systems and make them understandable to laypeople. As Jennifer Light argues, games during this period played a central role in spreading systems thinking into the popular consciousness. "Scholars must attend to the history of games," Light writes. "Returning these artifacts to a more prominent place in accounts of the period offers insight into the ways that individuals and institutions in the fields that made systems thinking central to their technical work simultaneously conceptualized lay knowledge about systems as fundamental to citizenship in a democracy."[41]

From the beginning, computer games had pedagogical and affective dimensions that were premised in part on making what was invisible or difficult to see—how they work—visible and sensible through representational and interactive moving images on a screen. The history of video games usually begins with two early computer games: *Tennis for Two* (Brookhaven National Laboratory, 1958) and *Spacewar!* (MIT, 1961). In the formalization of game studies as a discipline in recent years, these early computer games have been named important technological and cultural precursors to the video game boom of the 1970s. These games, developed in the computer research labs of the postwar military–industrial complex, are understood as establishing some of the formal grammar of video game mechanics and interaction and also as products of a celebrated hacker mystique that prefigured the personal computer. While that is true, they might actually be more significant for their affective and pedagogical functions at a time when computers where still widely feared by the general public. Both games were used in demonstrations for laypeople visiting

Early computer games like *Spacewar!* were affective interfaces between computational processes and users. Courtesy of the Computer History Museum.

the labs because the games made visible—in some sense externalized and made available to the senses—the otherwise invisible and incomprehensible processes of the mistrusted machines.

Video games, as demonstration programs, were computers' ambassadors to the general public; they were used to make the machines seem friendly and accessible in an era before they became "personal." In this way, at their origin, video games were affective interfaces between computational processes and users, proxy systems that metaphorized in real time the cybernetic imbrication of mind, body, and code. The early gamelike computer programs *Tennis for Two* and *Spacewar!* gave laypeople a sense of agency and communion with machines they did not entirely understand and even sometimes feared. These programs provided a way for people to feel the power of computers—to apprehend that power through playful interaction—without needing to comprehend fully how they worked. This attitude is now pervasive in relation to our everyday engagement with complex digital systems.

The relevance of Tomkins's theory of the affects for thinking about video games is multifaceted. First, like Sedgwick and Frank, Lisa Cartwright, Richard Grusin, and others who have recently taken up Tomkins's work, I find in his theory a notion of the self as constantly formed and re-formed through daily, ordinary interpersonal and intrapersonal transmissions of feeling and sensation.[42] It is important to ask, then, as I do in this book, what role video games play in this forming and re-forming as part of our contemporary sensorium. Second, the central premise of Tomkins's theory is that the biological and social apparatuses of affect are "coassembled" with cognition. *What we know* is inextricably bound up with not only how we know it but also how we feel about how and what we know. This idea is useful for thinking about the epistemological and pedagogical aspects of video games. It is also useful for thinking about our orientations toward video games and how affects shape and are shaped by these orientations. Finally, and underpinning the previous two points, Tomkins's theory is invested in computational metaphors that yoke the biological and the social together into a complex system. As such, it offers us a way to think both about how theory is always a product of its historical context and about how this, as its orientation, has affordances and limitations.

As Patrick Jagoda argues, "To achieve a sense of videogames as a unique art form requires a more intimate understanding of the new sensorium that they open up." He continues, "This sensorium is not simply a dimension of the history *of* videogames; it is also productive of a broader experience of history as such that is nonetheless grounded in digital game technologies and their formal expressions dating back to the mid-twentieth century."[43] Historicizing video games alongside affect theory through the cybernetic fold points to how both are premised on historically specific ideas about the relationship between humans and computers, but it also calls attention to the limitations of both these premises and history as a methodology. Just as video games can give us access to the historical grounding of our current sensorium, as Jagoda contends, they can also give us access to the limitations of this historical grounding. Groping around in the dark in *Kentucky Route Zero*, we can begin to feel history and its limits.

Reorienting Game Studies

The feelings of disorientation in *Kentucky Route Zero* are evoked by the enfolding of space and time, but they also circulate through the game's self-reflexive humor. As Nick Montfort and others argue, humor in adventure games is usually premised on the limitations of programming.[44] Humor around the limits of computer intelligence has been present in adventure games since Crowther wrote *Colossal Cave Adventure*. In that game's first version, many words and instructions entered by players tested the limited vocabulary of the game and resulted in unintentionally funny responses from the computer. Later iterations of *Colossal Cave Adventure* were better prepared to account for all the possible ways players might enter commands that exceeded the program's limited vocabulary, always redirecting players to refine their responses to align with the game's limitations. Influenced by conversation bots like ELIZA, computer programs meant to mimic human conversation, Crowther and Don Woods (who significantly revised and added to the game) designed *Colossal Cave Adventure* to give joking responses to particular commands. For example, when faced with a dragon, if the player enters the command "kill dragon," the program responds, "With what? Your bare hands?" This response signals the fact that many actions in adventure games require the naming of items that the players have picked up along the way to achieve particular results. In this example, however, what seems to point toward a limitation—human language interfacing with computer language—is also the clever setup for another humorous exchange. If the player persists and responds "yes" to the program's mocking query, the program's response is "Congratulations! You have just vanquished a dragon with your bare hands. (Unbelievable, isn't it?)"

In early adventure games, humor appeared in moments when the limitations of the software were revealed to players. Entering textual commands that the program did not recognize would sometimes result in terse or ungrammatical responses. Krista Bonello Rutter Giappone argues that the humor of adventure games is related to Henri Bergson's idea of the comic arising in moments of "mechanical inelasticity"—the folly of a person who, through a kind of machinelike repetition of behavior, cannot adjust to changed circumstances.[45] However, it is in the inelasticity of mechanical and digital processes—the limitations of programming and players' enjoyable interaction with those limitations—where adventure games appear comedic. As such, humor is used in the genre to "recapture" the moment of limitation, tucking the seam back into the game's narrative as a sort of meta-medial joke about the relationship between the vastness and fantasies of the imaginary world and the limitations of code. As Giappone argues, the player seeks out these moments of limitation in the game's programming as parodic deviations where she is pulled out of the fiction for a moment. The joke is on the machine, and this points to the ways games can affectively transform into a pleasurable generic convention the otherwise frustrating experience of dealing with the limitations of computational systems.

In *Kentucky Route Zero,* humor through limitation tends to emerge at moments when the player is asked to interact with a machine interface in the game's narrative. For example, at the very beginning of the game, when Conway uses the gas station computer to locate Márquez, the player is faced with three choices: Messages, Addresses, and Games. Mirroring the game's own textual interface, Conway must interact with the operating system through a command-line interface. This *mise en abyme* moment, when the character's actions in the story mechanically merge with the player's interaction with the game's textual interface and the choices it provides, is the context for the game's first self-reflexive joke. Clicking on "Messages" or "Addresses" provides the player with further clues about the story. However, clicking on "Games" produces the response " 'Games' is not real," an echo of the ungrammatical responses sometimes produced by early adventure game software when it did not recognize a player's input. Of course, the game's human authors, not a software program, generated this line of text as intentionally ungrammatical. Besides serving as a wink to the player about the history of adventure games, " 'Games' is not real" points back to the preceding sequence in which Conway encounters the spectral players in basement. Who or what "is not real" is called into question again. This moment also sets up a series of references later in the game to the relationships among computer interfaces, the limitations of design, and how we access history.

Playing a contemporary adventure game like *Kentucky Route Zero* on a Mac now offers an experience of design and technological limitations that is very different from what the players of *Colossal Cave Adventure* encountered in the 1970s. What does it mean to expose the limitations of computers when these limitations are no longer structured or experienced in the same way? In *Kentucky Route Zero,* instead of typing in textual commands, the player clicks on dialogue or action choices offered by the program. In this way, the game resembles the choose-your-own-adventure novels that are one of its cultural touchstones. Each choice represents, at least theoretically, a different path taken in the game. *Kentucky Route Zero* continues the tradition of self-reflexive humor not to "recapture" exposed

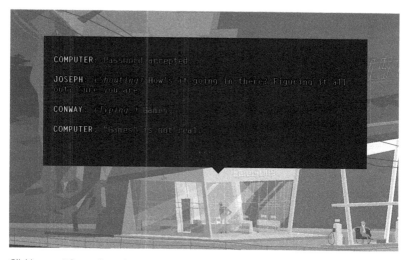

Clicking on "Games" on the gas station computer in *Kentucky Route Zero* produces the response " 'Games' is not real," a reference to the ungrammatical responses sometimes produced by early adventure game software.

seams of programming but rather to make these seams even more explicit. The game offers the appearance of choice while also limiting the range of player interactions, thus making the self-reflexive humor of the adventure game genre doubly reflexive.

In that the game's software offers prewritten text as choices, the player's task is less difficult than that of a player in a traditional text adventure game. The *Kentucky Route Zero* player's choices do not involve a process of understanding the limitations of the game's vocabulary; rather, the player must try to predict what effect, if any, particular choices will have on the game's narrative. The player quickly learns that the choices offer the illusion of many branching paths, but in reality each has only minor effects on how the game's story unfolds. This is reflected in the types of choices provided—they tend to be tonal rather than operational. For example, the first decision to be made is what to name Conway's dog. Joseph asks, "Did I hear a dog? What's your dog's name?" The player is given three choices of how to respond:

CONWAY: His name is Homer.
CONWAY: Her name is Blue.
CONWAY: Just some dog; I don't know his name.

This decision is not about picking a particular narrative path, but rather about affective engagement with the story. What does the player *want* to name the dog? How does she want to gender it? How does this decision give a particular emotional tone to the player's experience of the game and affect her identification with the characters?

In *Kentucky Route Zero*, the humorous moments that reflect on the limitations of computational systems are not just comedic asides; rather, they are central to the game's theme of the limitations of systems of all types—language, networks, bureaucracies, families, histories—and those made most peripheral and vulnerable by them. At key moments in the game, such as the one with the computer in the gas station, the narration becomes explicitly self-reflexive. These moments are all built around the merging of the characters' and the player's interactions with interactive systems in ways that comment on how those systems structure engagement and encode memory. Like Tomkins's theory of the affects, the game's mechanics and narrative trouble clear distinctions between outside and inside, between surface and what lies below. As such, *Kentucky Route Zero* is about the perceptual limitations of how video game history has been told—as a teleological series of gamers, hackers, and platforms—and the ways these limitations have made it nearly impossible to see women in this history. It is through the self-reflexive humor of the game that we can see the mutually constitutive limitations of how game studies tends to treat "code" and "history" as different from and beyond the purview of representation, cutting them off from feminist critique.

The self-reflexive humor in the game is critical to an understanding of how *Kentucky Route Zero* provides a feminist reorientation of the dominant historical and theoretical approaches to studying video games. The humor provides a model through which to examine how the player's experience of the comedic limitations of software can also create a context through which to reconsider the limitations of approaches to video games that fetishize hardware and code, and the bodies and experiences these approaches tend to exclude, specifically women's. The humor of the game is formally bound up with the game's retelling of early video game history through the fictional figure of Lula Chamberlain.

As we grope around in the dark, we unravel the mystery of Chamberlain's career as an artist and her involvement with Xanadu. Sometime in the 1970s, we learn, Lula, Donald, and Joseph meet each other in graduate school and collaborate in the creation of a vast virtual word that mirrors their own lives. The program, Xanadu (a reference to Ted Nelson's alternative to the World Wide Web), runs on a computer with organic mold-based circuitry, a layering of the biological and the mechanical. They take their project into the caves of Kentucky, where they encounter "strangers"—skeletal automatons working for a shady distillery operation. A doomed love triangle forms, the project falters from sabotage by the strangers, and only Donald and some research assistants are left in the cave working on the project. Joseph returns to the surface (where we meet him at the beginning of the game at the gas station) and Lula goes to work as a clerk at the Bureau of Reclaimed Spaces along the underground Route Zero, where she struggles to continue to make her art.

Lula Chamberlain's name, like the names of many characters in the game, carries historical weight; it is a reference to the computer programmer William Chamberlain, who designed the computer-generated prose program Racter. By putting the mystery of Xanadu and Chamberlain at the center of the game, *Kentucky Route Zero* poses a series of questions for the player. First, are the particular video games and virtual worlds with which we interact the inevitable outcome of the best rising to the top (i.e., a simple reflection of players' desires and market forces), or are there other forces at play? Second, does technological "progress" necessarily make games better? And finally, how might the present state of video games be informed by a broader landscape of artistic and intellectual influences?

Lula's story is first presented in the mini-game *Limits and Demonstrations,* which was released between Acts I and II. Here, the title of the mini-game directly invokes *Kentucky Route Zero*'s interest in constraints and limitations. In this mini-game the player controls the characters Emily, Bob, and Ben—the three vanishing tabletop game players from the basement of Equus Oils—as they visit a gallery exhibition (set at least two decades after the time of *Kentucky Route Zero*'s narrative) that is a retrospective of Lula's career as an artist, focusing on works that were too large and complicated for their time. The wall text for this exhibition explains, "Galleries and museums balked at the scale, power requirements, and highly skilled labor involved in maintaining these works for display. Some of their debuts collapsed under the weight of logistics, only to be successfully executed much later." The game, its disorienting timeline turning in on itself once again, suggests a kind of Benjaminian view of history, where the past is made visible only through the lens of the present. The wall text goes on, "Just as they describe the outer limits of Chamberlain's range as an installation artist, the geographical edges and vertices of her itinerant home life, and the beginning and end of her distinguished career, the works on display here also trace the extremes of our capabilities and the frontiers of our patience as both viewers and exhibitors."

Stopping in front of a work labeled "Overdubbed Nam June Paik installation, in the style of Edward Packard, 1965, 1973, 1980," Emily and Bob ponder what it is. It appears as magnetic audiotape affixed to the wall of the gallery, with speakers on either side visually citing Paik's 1963 multimedia installation *Random Access*. The title also alludes to Packard's Choose Your Own Adventure series of novels, which were first published in the 1960s. Ben explains, "It's an installation by a different artist, made of audio tape, and then she took it and recorded over all the tape with her own sounds." The magnetic tape contains recordings made by Chamberlain that document the process of creating Xanadu. Ben demonstrates that it works by running the playback head over the tape. They agree to "start in the middle." In a bewilderingly complicated enfolding of formats and interfaces, the text-based interface of the game remediates an audio recording on magnetic tape that recounts the creation of a hypertext system, Xanadu, that was operated by rotating punch cards on a mainframe computer.

> LULA: I usually start here, in my home, whenever I am sketching out a new piece. I start just by looking around.
> My closet door is open and I can see a few sweaters and a dress I like . . .
> COMPUTER: To go downstairs, rotate eleven degrees and advance four inches. To think about dresses, rotate one-hundred-fifty degrees and advance fifteen inches.

The player takes the role of Emily in this sequence. Emily is given two choices: "Go downstairs" or "Think about dresses." Here, through Lula's recording, the player is positioned domestically—"I usually start here, in my home"—to reflect on the gendered orientation and infrastructure of the creative process. Technological innovation here is consciously feminized through the domestic setting, and the choice "Think about dresses" drives the point home. This moment of imaginary history, rendered through an imaginary interface (rotating punch cards), resonates with Nooney's description of Roberta Williams sitting at her kitchen table, using the back of some wrapping paper to sketch out ideas for what would

Emily, Ben, and Bob look at "Overdubbed Nam June Paik installation, in the style of Edward Packard, 1965, 1973, 1980" in *Limits and Demonstrations.*

become *Mystery House.*[46] Attempting to "add" Williams to tech history, we might align her kitchen table with the famous "garage" settings of heroic male hackers and tinkerers of the same era. Nooney argues, however, that these versions of history overlook the affective dimensions of the table and how the domestic setting was a gendered orientation that affected the games that Williams designed. She writes: "If we start with 'the table' and not with 'the game,' *Mystery House* is precisely the opposite of a 'dramatic fracture' with the past. Rather, it is about what was most everyday for Roberta Williams, what was not simply a context but a material instantiation, her self's most intimate horizon of possibility and imagination: home."[47] From this reorientation to history, the fact that the first graphical adventure game is set in a home, rather than in outer space or another fantastic realm, and that players are required to "utilize mostly everyday objects to reveal secrets that exist beneath and beyond the immediately visible architecture" takes on new significance.

Through its many historical, spatial, and technological disorientations, *Kentucky Route Zero* privileges the very importance of the epistemological function of orientation. Emily, Bob, and Ben agree to "start in the middle," a phrase that appears several times in different parts of the game. Starting in the middle, Wendy Chun argues, is perhaps the only way to undo new media mythologies of all-powerful software or the all-powerful user. In Chun's formulation, entering a story (or an intellectual debate) in medias res draws out the ways digital technologies require us to perceive and imagine that which we cannot see or know through direct sensory perception. "Rather than offering a smooth chronology, the past is introduced through flashbacks—interruptions of memory."[48]

We hear Lula's story again in Act III, when we finally interact with the program itself deep in the cave. The hybrid system of old computer parts and mold begins this time not with Lula's voice but with Crowther's and the opening line of *Colossal Cave Adventure*—a different interface, a different orientation. As with much of the game's reflexivity, this does not feel like an erasure but rather just another enfolding. In *Kentucky Route Zero*, everything is a potential storage format and a platform for recording over. The characters Shannon and Chamberlain remind us of this fact in their appropriation of and reframing of the bodies and histories of computer science. The corporeal and discursive, the analog and digital, are meaningfully entangled and, like magnetic tape, are sites onto which different histories and different affective scripts might be recorded.

The game allows us to imaginatively feel video game history and its limitations and, in doing so, affectively reframes that history in ways that permit a different feminist engagement with the medium—one that is not

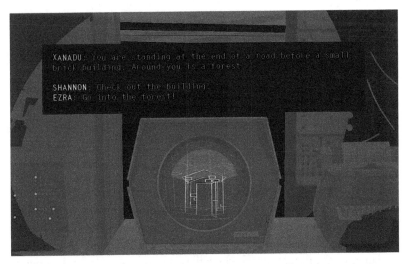

In the disorienting cave of *Kentucky Route Zero,* Xanadu remediates the opening lines of orientation from *Colossal Cave Adventure*: "You are standing at the end of a road before a small brick building. Around you is a forest."

solely about the representation of women in games but also about expos-
ing gender as an infrastructure in video game history and culture. Through
its speculative reimagining of early video game history and humorously
self-reflexive narrative and mechanics, the game creates an affective encounter with history. Between the simple point-and-click interactivity,
the characters and references that keep returning us to the significance of
cybernetics to twentieth-century thought, and the self-reflexive humor
of the adventure game genre, *Kentucky Route Zero* offers a complex medi-
tation on what the medium can and cannot do—broadening our conception of what video games have been and can be in culture.

Conclusion

Midway through the first season of *Halt and Catch Fire,* a television period
drama about a fictional company in Texas in the early 1980s that is attempting to build a personal computer to compete with IBM, the show's writers
use *Colossal Cave Adventure* to cleverly link code, hardware, and gender in
an episode's multiple story lines. One story line hinges on the company's
securing the cooperation of a Japanese firm to provide discounted LCD
screens for the new PC. The other main story line concerns Cameron, the
punk prodigy coder whose ideas are routinely dismissed by her employers
because she is a "girl" in a hypermasculine field, as she strategizes a way
to become the programming manager for the project. Cameron uploads
Colossal Cave Adventure onto the company's mainframe to derail the programmers' work. In a delightful sequence, a dozen programmers, all men,
play the game long into the night, calling to each other over their cubicle
dividers in frustration and to offer tips. When their boss walks onto the
floor and orders them into his office, we imagine that he is angry about
their obsession with a game when they should be working. The tension
is broken with laughter when he unexpectedly turns to them, while gesturing in exasperation toward his computer monitor, and asks, "How the
hell do I get out of this godforsaken cave?" The following morning, Cameron reveals that this was a test to identify the savviest coders. In her mind,
these are the ones who did not play by the explicit rules, but rather sought
out the game's limitations and hidden shortcuts. Those who "cheated," she
explains, are the programmers who will write the best code for the operating system they are developing. The use of the game was actually a power
play by a woman struggling to demonstrate her worth among those who

are consistently doubtful that she belongs in their world. The links among the screen, the code, and the game merge in the following episode when Cameron has the epiphany—one rejected by her employers—that the PC's operating system should address the user as a text adventure game does, as a humorous interlocutor.

Our contemporary moment of misogynist attacks on women in gaming culture (brought to wider attention through #Gamergate) is directly related to the way women have figured in the history of the computer industry. The video game at the center of this story in *Halt and Catch Fire* offers a moment of play and levity in an otherwise highly fraught landscape where the contributions of women have been rendered invisible by an infrastructure that privileges certain narratives about what a computer is, who can be a programmer, and what makes a video game good. In the same episode, Cameron's coworkers laugh at her when she announces that she wants to name the operating system Lovelace. They think she is referring to the pornographic film actress Linda Lovelace. They stare blankly when she tells them that the name is actually an homage to Ada Lovelace, "as in the very first computer programmer." Cameron is a killjoy in much the same way Zoë Quinn, Anita Sarkeesian, Brianna Wu, and others are figured by some gamers as humorless "social justice warriors" bent on taking all the fun out of video games. By the beginning of *Halt and Catch Fire*'s second season, Cameron has joined forces with another woman to start a video game company.

This vignette from a television show is useful because, in fictitiously staging the origins of the personal computer revolution, it sheds some light on the ways gender, history, and video games cautiously orbit each other in ways that speak to the contemporary moment. In 1995, when Sedgwick and Frank published their Tomkins reader, the Internet had only recently been popularized through the browser-based hypertext system the World Wide Web. Digital media studies was a nascent field, and there was a renewed interest in cybernetics. Yet few digital media scholars took cues from Sedgwick and Frank's theoretical intervention. If anything, 1990s digital media studies reinforced the digital/analog binary. Game studies, also emerging at this time, inherited these trends. As discussed at greater length in the following chapter, the vestiges of the narratology versus ludology debates in the field have resulted in curious divisions between approaches atomized around the text (the realm of representation and traditional literary, film, and media analyses), those centered on the player (social scientific theories of play and reception), and those focused on the code (analyzing the "deeper" structures of programming). In game studies, it is almost as if poststructuralism, or cultural studies for that matter, never happened. Thinking about game studies through the lessons of the cybernetic fold and Tomkins's theory of the affects reorients us to how much the field can gain from some groping in the dark, from being less sure about the autonomy of these approaches or even especially about what or whom exactly they purport to study.

Affect theory and game studies must hold on to what is disorienting about the fold—historically and theoretically—without throwing out bodies and their orientations entirely. "Are we inside or outside right now?" "Start in the middle." "Think about dresses." "Caked with mud like chocolate frosting." Like Patricia Crowther heaving her exhausted body through the dark and narrow crevices of the cave, we must hold on to the body, its corporeal and discursive properties, and its affordances and limitations. Affect does not offer us a way out of language. Video games do not offer us a way out of the politics of representation. The fact that both so frequently claim these victories is a history worth rewriting.

CHAPTER 2

Touching Games

Look. Listen. Touch.

—*Superbrothers: Sword & Sworcery EP* (Capybara, 2011)

I N TOUCHING GAMES, what is it that we feel? Released in 2011 for mobile touchscreen devices, *Superbrothers: Sword & Sworcery EP* is a curious game. With its intentionally misspelled title, simple mechanics, and sarcastic, world-weary characters, the game does not suggest the possibility of evoking strong feelings. Yet this independently designed and published game became a surprise hit and the subject of critical and popular discussion about its strangely moving world. In a blog post titled "Inconsolable," the game's creators note with some surprise the many letters they receive from players expressing the ways they have been profoundly moved by *Sword & Sworcery*:

> Some have written in to speak about their love and affection for the project and its world, others have written about how the project has been a source of inspiration and motivation for them, which is encouraging for us to hear. We've heard about how the experience of playing *Sword & Sworcery* brought some people together with their child, parent or loved one. A few have written in to talk about how the project moved them, how it shook them, or how it helped them through a tough time, an illness, a loss.[1]

Being touched by a video game, I argue in this chapter, is directly related to the affective circuits that touching opens up between representation, screens, code, and bodies. The sense of touch has served as a philosophical aporia from Aristotle through Jean-Luc Nancy, Maurice Merleau-Ponty, and Jacques Derrida. Touch, specifically one hand grasping another, is central to Merleau-Ponty's understanding of the subject as always inextricably sensing and sensible.[2] More recently, touch has been taken up in affect theory through the work of Sara Ahmed. Touch is integral to how Ahmed identifies affect in the "zones of contact" between bodies, objects, ideas, and other senses that make impressions and otherwise give shape to them.[3] Touching in *Sword & Sworcery* creates an affective assemblage that involves the player's body and its sensual capacities, the screen and its sensual capacities, and the code of the game. The screen is a zone of contact that brings all of these things into relation with one another in ways that give shape and meaning to them through their effects on each other.

In this chapter, I explore the affect of the ludic interface by taking touch and screens seriously in relation to what they allow us to feel about video games and more broadly about computer mediation. An analysis of the role of touch and being touched in *The Empathy Machine* (merritt k, 2013), *Dys4ia* (Anna Anthropy, 2012), and *Sword & Sworcery* helps us to feel our way around these interwoven strands. To touch and to be touched is the capacity that affect describes. What kind of encounter is it when we touch the screen of a digital device? My initial answer to this question is that it is an intimate encounter—meaning that, like all touching, it simultaneously designates self and other *and* the confusion between these designations. My second answer to this question is that it is a kind of everyday entanglement that points to the ways bodies, devices, images, and code are enmeshed in basic ways. To analyze this entanglement we need to avoid the surface/depth or revealed/hidden binaries of textual analysis.

Intimacy, the importance of touch and proximity, is a key factor in how we understand digital interactivity. You may ask, is *intimate* the right word? The larger context of sociality—of being in relation—through networked computers can also be a context for disconnection or, at best, touching at a distance. Common sense associates the digital age with faux intimacy and alienation. Advertisements for the techno-dystopian television series *Black Mirror* featured an image of a creepily distorted reflection of a face as it peers into the shattered screen of an iPhone. The series—which imagines a near future in which people are frighteningly absorbed in their digital devices—figures the screens of such devices as murky reflective pools into which we obsessively gaze, not knowing, or even caring, about what lies beneath. *Black Mirror* taps into a popular sentiment about mobile devices like smartphones and tablets: that for all their allure and usefulness, they may be aiding and abetting humanity's narcissistic and sadistic tendencies. The idea of our digital screens as black mirrors emphasizes the intimacy of our digital devices, the way we hold them close and gaze intently at them. This is the realm of analysis that privileges our use, our engagement, the subjective and the personal. The screen as black mirror also suggests something dark and hidden from view, that which is imagined as existing somewhere behind or beneath the screen. What this perspective asks for is an engagement with what the glossy screen obscures, what our absorption in the image conceals from view—primarily the way code is perceived to function as the invisible substrate to the image. This is the bifurcation that separates representational and phenomenological modes of digital media analysis from computational modes that explore what the screens appear to distract us from.

Through technological, aesthetic, and historical affordances, video games offer us ways of being with and feeling machines. A phone resting in the palm of the hand beneath a hovering thumb is a different bodily comportment than shoulders hunched forward, eyes moving across text and images on the screen, a hand gently resting on a mouse. These attitudes in relation to computers are not restricted to gaming of course; they now constitute our everyday bodily engagement with all kinds of media, from work to leisure. How do we parse the shifting experiences, subtle and not so, across bodies, time, and machines, from leaning on the laminated particleboard of an arcade cabinet to holding an Xbox controller in both hands to tapping the screen of an iPad? This question appears throughout this book, from several angles. This is in part a theoretical inquiry into the entanglement of bodies and machines, what these entanglements feel like, and how they register at the interface of computation, representation, and proprioception. It is also a historical inquiry into how and why some of these entanglements came to pass, and how their histories inform how they feel in the present. Being touched, in the sense of being moved emotionally, signals the very real experience of something immaterial pressing itself on our emotional state and changing it, and the capacity for us to touch someone or something else in a similar way. While the vestiges of cybernetics in affect theory discussed in chapter 1 can certainly be seen as a limiting structure for complex biological and social processes, such a structure is actually quite useful for describing a dynamic of sensory intersubjectivity among humans, digital devices, and computational processes.

What can we learn about our affective orientation to the world as we live it through the entanglement of embodied and mechanical circuits? How might we think through the digital gestures of tapping, pinching, and swiping as participating in circuits of affective engagement with digital media and their computational and representational systems? And how might this open up new critical pathways for feminist and queer analyses of video games? By taking the surface seriously—the screen both as a material object with specific historical and technological affordances and as a site of the everyday entanglement of bodies and representation—in this chapter I explore the larger question of how video games "touch" us. The previous chapter was about how video games might affectively reorient us toward history and how particular histories of video games create limits to what we can know and feel in the present. This chapter is about how video games might put us into contact with the sensual properties of code and also about that which gets figured as matter available to touch and that which is figured as immaterial and out of reach.

Touching the Screen

We now touch many different types of images in our daily lives, and often our touch is what instigates changes in them. Mostly, this image touching happens on mobile digital devices equipped with capacitive touchscreens. Touching images—gently tapping the "play" icon on a YouTube video, pinching or reverse pinching an image to make it appear larger or smaller on the screen, swiping right to indicate desire in Tinder, tapping the screen to select a reaction icon on a Facebook post—causes their movement on-screen and through networks, as well as the movement and actions of invisible digital processes. These are not simply interactions between our feelings, as properties of our bodies, and the deep structures of code as the property of networked computers. Rather, our bodies, their capacities to feel and to represent emotions—desire, love, anger, boredom—are necessarily enmeshed with the interfaces, their representations, and code itself as a form of representation. The touchscreen as a surface evokes the impasses of surface and depth, representation and computation, connection and disconnection.

Laura Marks's concept of "haptic visuality" seems to get at some of these contemporary interactions among touch, the senses, and imagery. In an attempt to redress the ocularcentrism of film and visual studies, in which the eye remains distanced from both the image and the rest of the body, Marks reattunes us to a kind of sensory wholeness in the perception of images. Haptic visuality occurs, according to Marks, when moving images deflect the optical gaze through a kind a synesthesia and the eye is drawn into the image's textures, patterns, and colors, producing a multisensory engagement with the film or video. In Marks's work, the transference of feeling from image to the eye and from the eye to the rest of the body is a process of both "dynamic subjectivity" and transcultural connectivity between viewer and object that occurs outside the boundaries of symbolic representation.[4] But the eye still leads in her formulation, in that it transfers the sensory information taken in optically to other sensory registers. The limitation of haptic visuality for the kinds of screen encounters I am discussing is that it tells us little about direct touching of images, and it does not account for the ways in which this touching affects the images and vice versa.

In the field of human–computer interaction (HCI), direct gestural manipulation of on-screen information is assumed to be more "natural" or "intuitive" than other types of common digital interfaces. This idea goes beyond HCI discourse and circulates as cute stories of iPad-crazy toddlers who go around swiping nontouchscreen objects, and kittens and other animals that play touchscreen video games. And, as Elizabeth Ellcessor argues, this is the "you already know how to use it" marketing discourse that Apple has used for its mobile devices.[5] The sense that touchscreens are intuitive presents them as a natural outcome of technological progress, occluding their longer history of uneven development. Additionally, as Ellcessor demonstrates, this discourse assumes an ideal or preferred user and is destabilized by bodies that do not fit this ideal. The disabled body in particular, Ellcessor notes, "forces consideration of specific arrangements of bodies, technology, culture, and power."[6] The touchscreen, like all screens, is not simply a technology but also a technosocial assemblage.

When screens became regular features of digital computers in the 1960s, they reoriented the user's bodily, perceptual, and affective relationship to the machines. Besides becoming the primary location for the output of information, the screen was central to the process of making computers "personal." The small size of the original cathode-ray tube (CRT) screens addressed single users before there was such a thing as a personal computer. And their resemblance to television consoles provided familiarity for the new machines. The computer's provenance in the military–industrial complex of World War II, application to Cold War intelligence, and remote complexity had made it an understandable target for a lot of negative affect in the popular imagination, spawning conspiracy theories, distrust of authority, and science fiction visions of mind control. CRT screens were a step in the direction of visualizing the abstract data being processed and making the labor of computers more apprehensible. In 1961, it was the screen on the new PDP-1 that inspired Steven Russell, Wayne Wiitanen, and J. M. Graetz at MIT to create *Spacewar!* as a visual demonstration of the computer's enigmatic power. Graetz recounts the context for the development of the game:

> When computers were still marvels, people would flock to watch them at work whenever the opportunity arose. They were usually disappointed. Whirring tapes and clattering card readers can hold one's interest for only so long. . . . The main frame, which did all the marvelous work, just sat there. There was nothing to *see*.
>
> On the other hand, something is always happening on a TV screen, which is why people stare at them for hours.[7]

Digital computers were not originally optical technologies, but CRT screens, and what they made visible and sensible about computers, played a key role in their cultural expansion. The screen appeared as the friendly and expressive face of the computer.

The computer screen as a ubiquitous interface for both input and output is a relatively new development. Although the capacitive touchscreen technology used in contemporary mobile digital devices was initially developed at CERN in the early 1970s, the touchscreen has only quite recently replaced the keyboard and mouse as the dominant digital interface. In the 1960s, HCI researchers developed various interfaces to make computers more accessible by focusing on a user's ability to manipulate by hand objects represented on a screen. In 1962, Ivan Sutherland's graduate work at MIT yielded Sketchpad, a program that is recognized as foundational in the development of the computer screen as a responsive surface. Sketchpad allowed a user to input data by touching the computer screen with a handheld light pen. Quickly, others sought to remove the mediating device and put hands in direct contact with screens. Interface design in the 1960s and 1970s established a relationship between the hand and what the eye sees represented on a screen. This desire to put hands in direct contact with screens persisted, at least iconographically, in the graphical user interface that designated pointer location on the screen with an image of a pointing hand. Even through the reign of the mouse and keyboard, hands on screens remained the fundamental paradigm and ideal of embodiment for digital interactivity.

The introduction of the computer screen as interface preceded the slightly later emergence of "geological" metaphors in computer science that rendered code as part of a layered deeper structure beneath the screen. Rita Raley traces this notion of code as "deep" to the late 1960s.[8] Through the 1950s and most of the 1960s, Raley contends, those working with computers did not use spatial metaphors to describe the relationships among code, software, and hardware. The idea of code as a "deep structure" began to emerge only as the relationship between software and interface design demanded a distinction between "machine languages" and "user languages." Raley revisits this history to explore the present tension between the idea that code is something deep and inscrutable (that it exists outside the realm of signification) and the coexisting idea that code has a logical structure that can be mapped and apprehended (that it is one language among many).

In the 1960s and 1970s, HCI discourse established the screen as a sensing interface between the computer's data and the data input by the user. The screen became the face of the computer and, as a result, was central in technological and discursive processes that figured the user's body as proximate only to a "surface effect" of a deeper obscured structure of information. And though it would take a couple more decades before the touchscreen would become commercially viable as the dominant digital interface, a particular dynamic involving hands, eyes, screens, and code had been established.

Game Feel

We have a very limited vocabulary to describe the difference between controlling video game action through touching images directly with our fingers and controlling the action through indirect controllers, keyboards, and pointers. How can we understand the appeal of using our fingers to pull a wingless cartoon bird across a mobile phone screen? Stroking the screen with just the right pressure and speed, and at just the right angle, we pull the slingshot taut and launch the bird toward the annoyingly happy pigs. How do we describe the peculiar satisfaction of watching a row of glistening candies burst into points after being matched in a row beneath our hovering hand? In recent years, the games *Angry Birds* and *Candy*

Crush Saga have transformed the video game industry and also how we use our mobile devices. Part of their appeal lies in their game mechanics and imagery, which heighten the feelings of connection between bodies and computational processes through the touchscreen interface. Through design, such games make us feel as if our fingers have the capacity to attract and *pull to the surface* the ephemeral objects of programming, rendering them fully available to our sensual world, with some of the properties of physical game pieces but also with the will and obstinacy of autonomous animate objects.

Game designers talk about "game feel" as a quality intentional in their designs. Crafting a game that a player will find challenging but not overly frustrating, visually compelling, and narratively satisfying, with mechanics and a game structure that produce the right amount of tension and gratification, is a hugely complex endeavor, and game companies have invested a great deal of resources in trying to figure out how to meet all these criteria. In his book on just this subject, game designer Steve Swink locates the crux of the issue in game mechanics. Swink uses words like "floaty," "tight," and "responsive" to describe how the design of a game's mechanics creates an overarching feeling for the game.[9] Game designers like Swink privilege programming and design as the site where a game's experience is created and use player experiences to refine and control game feel as much as possible.

Game feel, however, as Swink defines it, cannot fully account for what it feels like to play a video game. The responsiveness of the controller and the physics of the game world are important but small parts of what makes up our aesthetic encounter with video games. The term *game affect* gets closer to addressing this broader phenomenon. Eugénie Shinkle links affect and video games through precognitive processes of proprioception—the way the body processes subtle sensory information from the environment to track its position and movement through space. Pleasure in a game is derived, according to Shinkle, through a complex interplay of the body in proximity to the gaming apparatus, the feel of the game controller in the hands, and the images, sounds, and other aesthetic information in the game. Shinkle writes, "The rush you get from a good game is a *subrational,* bodily thing, involving phenomenological or affective dimensions which cannot be programmed into a game."[10] Valerie Walkerdine, on the other hand, in her work on gender and video games, uses a psychoanalytic approach to understand a player's affective experience of a game. "Feeling one's way around a virtual space," she writes, "is not simply a kinetic, or indeed kinaesthetic, experience but an affective one." She continues, "In this sense, affect means we look at three aspects—the sensation, the ideation of that as pain, for example, and the defence against the pain which is the fantasy of pleasure—that pleasure is the fantasy of control or being somewhere or someone else."[11] Although they use different approaches, both Walkerdine and Shinkle get us closer to thinking about video games and bodies as engaged in an affective circuit. Still, because of their focus on players and on affect as part of subrational or unconscious processes, Walkerdine and Shinkle provide little in the way of actual analysis of video games, their different affordances, or their affective specificities. Instead, they are interested in how video games, on the whole, might be remapping our affective lives. Shinkle engages with Brian Massumi's notion of affect as a virtual potentiality in that affect, as a semiautonomous force distinct from emotions, has the capacity to impinge on and rewire bodies to form new ways of being in the world. As Shinkle puts it, video games as relatively new technologies "have the potential to reorganize sensory priorities, to augment the senses and to modify the relationships between them."[12] Similarly, Walkerdine builds on Mark Hansen's idea that in our encounters with digital art, our bodies, mirroring computational structures, become "processors of information."[13] Following this, Walkerdine links her understanding of how affect works through video games to how subjectivity is formed by them, arguing that "thinking about the centrality of affect to new media is absolutely essential to our understanding of the pleasures afforded by video games and the ways that they make subjectivity possible."[14] As in Hansen's conception, then, the body and subjectivity—when faced with computational processes—seem capable only of taking up those processes, not of affecting them. This is not a meeting of different types of matter with their own unique properties and affordances; rather, it is a collapsing of language to render bodies and machines as operating in the same way.

For Swink and other game designers, game feel is something located deep in the programming. For Shinkle and Walkerdine, game affect is located in the body of the player, either in "deep" neurocognitive processes or in the unconscious. All figure the affective dimensions of games as somehow out of reach of observation and interpretation. Game studies is beginning to have meaningful conversations with affect theory, but these conversations face the impasses inherited from the "cybernetic fold"—the cleaving of the digital from the analog and the body from the mind. As I have demonstrated in chapter 1, thinking through this shared inheritance draws out how both game studies and affect theory are historically and theoretically imbricated. As such, both are prone to problematic computational and spatial metaphors that reinforce surface/depth, self/other, and inside/outside distinctions. In this chapter I further explore the ways surface/depth distinctions shape both fields of study and limit what they can say to each other. In game studies this is evident in the tendency to treat the images on the screen as merely expressions of a deeper and more meaningful system, one that is difficult to access and requires specialized knowledge to understand. We are told that our analytical tools—calibrated for the language of representation—no longer apply to the world of simulation. In reconsidering the screen as a site of representation and as an affective interface for the meaningful assemblage of technologies and flesh, I am rejecting this tendency. The rendering of the screen as surface expression of a deeper structure, I argue, does not actually offer a new theoretical tool for new media; rather, it adapts the ideological blindness of a previous era's surface/depth models of consciousness to computational media. In this chapter I demonstrate how the turn away from the screen in game studies leaves us without a critical language to describe and analyze the complex assemblage of identity, ideology, and subjectivity that pulses through video games as representations. I hold the turn away from representation in game studies alongside a similar tendency in Deleuzian affect theory to render "feelings" as merely affect (the deeper force) called up into subjective representation. Here, in this kind of affect theory, the surface/depth model renders affect as that which is out of reach and only indirectly accessible to analysis. In holding these similar intellectual moves together, I propose a model that attends to surfaces—of screens and bodies—not as sites of secondary expression but rather as crucial sites of touch and entanglement, where representation still matters and representation *is matter*.

Surface Effects

Midway through *The Empathy Machine,* merritt k's text-based video game about using computers to experience different genders, the game asks the player to put a hand on the screen. It is an odd request because the game is not designed for touchscreens; rather, the player advances by using a mouse to click on highlighted words. The game unfolds through text prompts that position the player as interlocutor with an artificial intelligence program designed, as it states, to digitally simulate empathy. *The Empathy Machine* is k's response to the development of virtual reality (VR) programs that allow people wearing VR headsets to try on bodies of sexes different from their own.[15] As a transgender woman, k is understandably critical of this very premise that the complex lived experience of gender—cis, trans, or otherwise—can be simulated through a VR program that allows a man to look down and see a female body in place of his own. It is significant that one of the primary goals of VR technologies is to do away with the screen, or at least the sense of a screen as a mediating interface, altogether. Here, however, the technological simplicity of k's game and its insistence on putting the player in direct contact with a computer screen are part of the designer's critique of the normative assumptions that continue to shore up the fetishization of simulation and immersion in digital media. Whether or not the player actually touches the screen makes no difference to the game's conclusion, which proposes that empathy and embodied experiences of gender are more complex than computer simulations. The game asks, "Is empathy really about trying to inhabit other people's experiences, or is it more about simultaneously holding difference and sameness without subsuming either?"

Game studies suffers from a kind of sensory deficit that is different from ocularcentrism. It can account for the physical presence of hands on interfaces, but not for what this has to do with the multisensory perception of images. In carving out the field of game studies, scholars have moved away from the screen and its images. Action-based theories and proceduralism create a circuit of meaning between the actions of the player's body—say, thumbs on a controller or hands on a mouse—and the actions of the mostly invisible digital and mechanical processes of the game's programming. The feel of the game controller in the hands become clearer in this approach, but the screen and all it conveys go strangely missing. The screen, what it obscures and reveals, is a curious problem for game studies. The field tends to emphasize "rules," specifically algorithms, as the site where video games' expressive potential can most meaningfully be traced. During the emergence of game studies in the early part of this century, some scholars were eager to distinguish the field from literary or media studies. Starting from the premise that rules are essential to how all games make meaning, video game scholars understandably built on the homology between "rules" and digital structures like algorithms (looking

please put your hand on the screen

y / n

The Empathy Machine invites the player to touch the screen as part of its critique of the fetishization of immersion in digital media.

past other rule-based systems, such as narrative, perspective, montage, and language) to form a foundation for video game studies in nonvisual computational rule-based processes. For example, in one of the seminal articles establishing this "ludological" approach to video games, Espen Aarseth argues that when we play our attention is elsewhere, on the rules and objectives of the game. "The dimensions of Lara Croft's body . . . are irrelevant to me as a player," writes Aarseth, referring to the game *Tomb Raider.* "When I play, *I don't even see her body,* but see through it and past it."[16] In Aarseth's formulation, the video game screen and its images are something that the player *must look through and beyond* in order to access what the game is really about: the successful manipulation of algorithms to win. Aarseth and others contend that a game's ideal theorist is one who, like its ideal player, can see past the screen and its images to an imagined deeper algorithmic structure. This approach in game studies might seem akin to what some visual culture scholars have recently suggested all contemporary media critics should do: look beyond simply what is visible in media into the interfaces, platforms, networks, and code to understand fully how they convey meaning. And, indeed, some early influential work in game studies put this field into a productive and necessary conversation with emerging developments in digital media studies, such as code and software studies and philosophies of new media that emphasized the computational structures and protocols of media. In game studies, however, looking beyond the visual has a slightly different intellectual history and different theoretical stakes that are not simply about supplementing visual analysis with code studies but rather advocate for taking a sharp turn away from the visual altogether.

The turn away from the visual in game studies is often premised on the idea that, as computational media, video games are different from so-called optical media, such as photography and film, and so require a different analytical starting place. In *Gaming: Essays on Algorithmic Culture*, for example, Alexander Galloway writes, "If photographs are images, and films are moving images, then *video games are actions*. Let this be word one for video game theory."[17] Galloway argues that games are essentially about *doing* and, as such, are intrinsically different from image- or text-based culture. In this highly influential book, Galloway defines the "work" of the video game as existing in an invisible space of interaction between designed software and user activity. Galloway's conception of action in video games relies on his reduction of action in other media, like film, to the fabrication of the works, leaving out, for example, the mental and embodied actions of film reception and interpretation. Similarly, in his later work *The Interface Effect*, Galloway continues this call for a move away from the visual in digital media theory more generally. To understand the relationship between ideology and software, he argues, "it is more valuable to separate . . . [the] visual and machinic aspects in mutually distinct struggle" because this separation mirrors a social separation between informatics-based models of ideology and "a slightly older one which takes as its prized aesthetic forms of the verbal narrative and the visible image."[18] This severing of the visual from the machinic allows Galloway to argue that the interface is merely an *effect* of something deeper and less sensually available to us. "While readily evident in things like screens and surfaces, the interface is ultimately something beyond the screen. It has only a superficial relationship to the surfaces of digital devices, those skins that beg to be touched."[19]

In another key text in video game theory, Ian Bogost calls not for a turn away from the visual specifically, but rather a turn away from all representation-based models of analysis. Bogost also bases his argument on the idea that video games' computational architecture necessitates the rejection of modes of analysis based on form, content, and signification that are overly focused on "surface effects."

With the advent of the von Neumann architecture, the software of
a computational system made a strategic break from the hardware,
allowing engineers to focus more on atomizing small computational problems even within the most complex of systems. In both
fields [poststructuralism and computer science], universalization
became an approach rather than an outcome, dispersing totalities of organization into less visible totalities of method. In this
sense, the continuing endurance of *surface effects* and *end products* in the arts, humanities, and technology industries demands
questioning.[20]

Here Bogost suggests that the atomization of computer architecture and
programming labor into smaller and smaller units necessitates an analytical approach that de-emphasizes an overarching system or clear outcome.
He offers instead "unit operations" and "procedural rhetoric" as categories
and modes of analysis that respect the complicated technological infrastructure of video games as software. For Bogost, representation in games
occurs primarily in their less visible capacity as *simulations*, which he defines as "a representation of a source system via a less complex system
that informs the user's understanding of the source system in a subjective
way."[21] The figuring of a "source system" here privileges an understanding
of computational media as layered entities through which we have only indirect and "subjective" access to some deeper, obscured, layer of meaning.
Bogost leaves unexamined the legacy of psychoanalytic theory expressed
in this formulation. Rather, he is more interested in suggesting a categori-
cal shift in theory away from representation and toward simulation. The
stake of this shift for Bogost and Galloway is a necessary turn away from
the image, representation, and the surface.

Recently, game studies scholars have pushed back a bit against this
turn by reasserting the value of textual analyses and a broader definition
of "rhetoric" in games and play.[22] Yet there remains the ludology versus
narratology hangover that requires game scholars to defensively make a
case for representation while also keeping it a realm of analysis separate
from mechanics and play. Computation/representation has become *the*
structuring binary for game studies. Attempting to define what video
games are, game studies textbooks continue to downplay the significance
of narrative and images, focusing instead on digital rule-based processes.
For example, in *An Introduction to Game Studies*, Frans Mäyrä makes a

strong case that studying games as "complex cultural systems" requires various methodological approaches, but his "main framework focuses on distinctions between gameplay and representational aspects of games," designating the latter as "core" and the former as "shell."[23] Rigorous video game analyses are not supposed to pay as much attention to narratives or images as to rules, mechanics, and underlying computational structures. As Mäyrä writes, undermining his own call for a "cultural dynamism" in game studies:

> Gameplay [the core] is what doesn't change when you change the surface: the rules.
>
> The gameplay isn't the entire experience of a game, but it's what makes it a game, what makes it *this* game....
>
> ... It's not the interface..., it's not the graphics and it's not the story. It's the part of the game that absolutely requires the player's participation.[24]

There is, indeed, something special about the computational architecture of video games that is central to how we experience them and make meaning through them. Video games both work as computational systems and can simulate other systemic relationships in compelling and thoughtful ways. A word-processing program is a rule-based system, but Microsoft Word and Apple's Pages do not intentionally create the context for meaningful reflection on their status as systems or on the behaviors and systemic relationships they model in their design. Video games, however, do this frequently and convincingly. Video games are meaningfully structured through rules and actions: press "A" to jump over the green pipe in *Super Mario Bros.*, align the blocks to build a Soviet wall in *Tetris,* choose to protect or kill the "little sisters" as you move through the failed objectivist society of *BioShock.* These rules and instructions enframe the player's interactions with the game and are the legible inscriptions of the rules and instructions—the code and algorithms—that allow the game to execute the player's commands and the software to communicate with the circuitry of the computer or console on which it is played.

Computational approaches have been extremely productive for game studies and have provided ways of identifying what makes video games a unique and culturally significant form of media. Still, there is something not wholly satisfying in thinking about video games first as procedural or algorithmic expressions. Video games compel us to act and to be acted upon through the procedures of their algorithmic structures, through the ways these structures are given representational form visually, narratively, and aurally, and through the designed intimacy of their interfaces and the contexts of our play. Contrary to Mäyrä's claim, gameplay *is* the interface in a significant way. On a basic level, video games demand that decisions be made at an interface where the player has access to a set of instructions (the rules, the narrative, the images) and the game has access to its own set of instructions (the code). They are a medium of *configuration*, as Galloway argues, in that what unites them is that they ask us to make choices in our use of them and to make choices about *how* we use them. We make choices and push buttons in games because of the complex interaction of various factors, one of which is how video games structure our feelings about those choices and actions through what appears on the screen. When we enjoy a video game (or not), how do we describe that experience and to what do we attribute our feelings? How does it feel to play a video game? What analytical frame best encapsulates the complex interplay of bodies, hardware, code, aesthetics, affect, and cognition? The computation/representation binary makes it difficult for game studies to address these questions.

This binary in game studies also privileges certain types of games and certain types of questions about games over others, reinforcing gendered hierarchies in gaming culture around issues such as what constitutes a video game and who its proper subjects are. Aarseth's statement about looking past Lara Croft's body resonates with a popular filmic representation of seeing and digital screens. In the film *The Matrix*, when Neo is first learning about the encoded simulation of reality, he asks Cypher what it is he sees in the bank of computer screens with encrypted data scrolling across them. Cypher responds, "You get used to it. I don't even see the code. All I see is blonde, brunette, redhead." *The Matrix* suggests the inverse of Aarseth's formulation of the digital screen; here we must look through and beyond the code to see the image. Yet in both formulations, images of women are the points, whether starting or ending points, for the masterful and probing digital gaze. The image of a woman is a transparent veil over the deeper computational structures, or code itself is a veil that, with some practice, might reveal a woman. Similarly, the computation/representation binary, dependent as it is on spatial metaphors of surface and depth, reestablishes the gendered spatial imaginary that Laura Mulvey identified in film more than forty years ago.[25] In game studies, the screen is the arresting and distracting feminine surface that obscures the deeper space of action and the masculine probing of code.

Surfacing Affects

The turn away from representation in game studies mirrors a similar move in affect theory. In his influential work, for example, Brian Massumi argues that cultural theorists' interest in "mediation" is what gets in the way of their understanding the sensing body in a postphenomenological fashion. Mediation, specifically the body as represented by and constituted through language, is an interface—a kind a surface effect—that shields theory from addressing what lies beneath it, an actual, sensing body. Writing about the dominant concerns of theory in the latter half of the twentieth century, Massumi argues:

> Culture occupied the gap between matter and systemic change, in the operation of mechanisms of "mediation." . . . Mediation, although inseparable from power, restored a kind of movement to the everyday. . . . The body was seen to be centrally involved in these everyday practices of resistance. But this thoroughly mediated body could only be a "discursive" body. . . . Make and unmake sense as they might, [mediated bodies] don't *sense*. . . . It was all about a subject without subjectivism: a subject "constructed" by external mechanisms. "The Subject."[26]

Massumi's critique of structuralism is not a wholesale rejection of its usefulness; rather, he argues that its entry point, language, creates a blind spot in that the subject can be grasped only through mediation—how language positions and fixes the subject in time and space. In Massumi's work, affect is the concept that moves theory away from the linguistic legacy of structuralism and poststructuralism. Affect, in this sense, shifts our attention away from the traps of the "discursive" subject to a consideration of a realm of action and "becoming" that is nonlinguistic, nonsymbolic, presubjective, and transpersonal. Massumi compares it to an echo bouncing between two walls: "An echo . . . cannot occur without a distance between surfaces." He continues: "With the body, the 'walls' are the sensory surfaces. . . . The emptiness or in-betweenness filled by experience

is the incorporeal dimension of the body."[27] Here "experience" is likened
to a kind of information or data that the body processes but that is not
entirely *of the body*. With his strangely hollow bodies, Massumi enacts the
familiar neo-Cartesian move that is inherited from cybernetics, and especially from theories of artificial intelligence that presume experience, intelligence, and consciousness can be conceptualized autonomously from
embodiment. This allows him to say a bit later, "Intensity [affect] is . . . a
nonconscious, never-to-be-conscious autonomic remainder. It is . . . disconnected from meaningful sequencing, from narration, as it is from vital
function . . . spreading over the generalized body surface like a lateral back-
wash from the function-meaning interloops that travel the vertical path
between head and heart."[28] This generalized body of sensing walls and vertical paths between the processing head and the emotional heart is filled
by experience and intensity, other terms for affect in Massumi's work. Affect here, then, serves neither a vital nor a symbolic function. Strangely, for
a theory that seeks to restore the sensing body to contemporary thought,
this model suggests not only that we can but also that we must make a distinction between "vital" and "symbolic" functions, between a generalized
surface and the deeper data of experience, and between how a body senses
and how a body makes sense.

Writing in response to the "affective turn" in theory, particularly the
work of Massumi, Ruth Leys critiques the way affect is used as a concept
to shift attention from analyses of how power makes itself visible and present to us as subjects:

> The whole point of the turn to affect by Massumi and like-minded
> cultural critics is thus to shift attention away from considerations
> of meaning or "ideology" or indeed representation to the subject's
> subpersonal material-affective responses, where, it is claimed,
> political and other influences do their real work. The disconnect
> between "ideology" and affect produces as one of its consequences
> a relative indifference to the role of ideas and beliefs in politics,
> culture, and art in favor of an "ontological" concern with different
> people's corporeal-affective reactions.[29]

Leys views Deleuzian strains of affect theory, in which the subject and representation are rendered as of secondary importance, as making an end
run around the necessary work of analyzing how power still very much

works through representation. Game studies and Deleuzian affect theory share a turn away from the visible that is premised on the proposal that new social and technological conditions require the abandonment of theory's structuralist hangover and the adoption, counterintuitively, of the surface/depth distinctions that precede structuralism and reinforce its blind spots. Bodies, all the ways their surfaces continually signify, and the ways they are mediated through representations are rendered by these discourses as merely "surface effects." Affect for Massumi is what gets bodies out of the grids of signification wrought by the legacies of structuralism, but his use of machine language to figure a sensing corporeal body still leaves us in the realm of the discursive—a realm his own theory of affect cannot examine. Specifically, in its particular focus on affect as virtual— something proximate to but autonomous from bodies and language—his theory is ill equipped to account for how the feelings of bodies come to matter in and through media in ways that are still quite bound to the symbolic power of representation. In turning away from mediation, and thus all the ways subjectivity and embodiment are not mutually distinct but entirely imbricated, a great deal of what goes under the name of affect theory seems content to leave unexamined the premise that a meaningful distinction can even be made between the corporeal and the discursive body.

Similarly, the computational approach in game studies not only disregards the screen and its images but, by extension, also renders any discussion of the links between representation and subjectivity in digital culture as retrograde throwbacks to cultural studies and identity politics. The tendency to cordon off the machinic from the visual (or informatics from representation) leads to strange formulations of what subjectivity, identity, and affect are. For example, toward the end of *The Interface Effect,* Galloway faults identity politics and the social movements of the twentieth century for focusing too much on the "liberation of affect." Technical structures of networks and digital protocols have perfected this process, he argues, by perfecting racial coding at the level of the algorithm. He writes:

> The crucial political question is now therefore not the liberation of affect, as it was for our forebears in the civil rights movement, the gay liberation movement, or the women's movement, in which the elevation of new subject positions, from out of the shadows of oppression, was paramount. The crucial question now is— somehow—the reverse. Not exactly the repression of affect, but

perhaps something close. Perhaps something like a politics of subtraction or a politics of disappearance. Perhaps the true digital politics of race, then, would require us not to "let it be," but *leave it be*.[30]

Similar to Ley in her critique of affect theory, I am troubled by the way game studies, and digital media studies more broadly, seems to want to move beyond the still-important question of representation by figuring a computational realm in which power works in ways that are detached from lived experience and, hence, legible only to those with the power to decipher them. The only recourse we have to combat this, according to Galloway, is to detach ourselves from politics and subjectivities based on identities, no matter how fragmented, hybrid, or transitory. Galloway's suggestion that we "leave be" categories like race, gender, class, and sexuality and assume the neutral subjectivity of the affectless "whatever" assumes that identities are purely choices that can be put on and thrown off at will—as when we configure an avatar from a menu of choices in a video game. His suggestion also assumes that affect is a quality that is under the control of the autonomous individual, rather than a relational force that (much like identity) is formed both socially and materially across bodies and objects. Affect, like subjectivity, is not something that can be either liberated or willed away. This model, as I argue throughout this book, makes it difficult, if not impossible, to see the ways computation and representation, like affect and cognition, bodies and intelligence, are completely intertwined. This entanglement can really be grasped only through mediation, the interface that permits one system to inform and shape the other.

The Empathy Machine says, "Please put your hand on the screen." If we do reach out to touch the screen as instructed, we feel a bit foolish as we press a hand awkwardly against the unresponsive surface. *The Empathy Machine* is cold and hard. The words of the text hang didactically against an empty black background. Yet the game does make us *feel something*. This is the paradox at the heart of the game and, I want to suggest, at the heart of the video game encounter. A computer program may not fully capture the complexity of what gender feels like, but that does not mean that it cannot create a circuit of feeling across screens, language, code, and bodies.

Video games can actually be quite profound empathy machines. Take, for instance, Anna Anthropy's game *Dys4ia*, an autobiographical account of her experience taking hormones as part of her gender transition process. Eschewing any appeal to an aesthetically realistic representation of Anthropy's experience, *Dys4ia* instead uses the graphics and mechanics of 1970s and 1980s arcade games to give a performative dimension to the feelings of frustration and shame and the thrill of small victories expressed in Anthropy's textual account of her experience. As we try to maneuver a series of blocks into place in a wall, à la *Tetris,* we are also reading Anthropy's first-person account of feeling like she could not get her body to fit either her own or the world's conception of what it should be. Playing an arcade game does not successfully simulate the feelings that Anthropy describes, but this is not the point. The game touches us through the disjuncture between the emotional complexity of Anthropy's autobiographical account and the simplicity of the graphics and mechanics. Anthropy knows that games signify not just through mechanics but also through images, stories, and the ways our bodies makes sense (or do not) across all of these planes of meaning. Anthropy, a game designer who grew up with video games, knows that the juvenile and gendered thrill of seeing pixelated breasts in a game can also be used, in a different narrative context, to express movingly the ways gender, bodies, sexuality, and technologies are imbricated. The mechanics of *Dys4ia,* however, do produce feelings beyond what the text or images alone could convey. The game's capacity to move us emotionally—to touch us—is directly linked to the movements it asks our bodies to make in relation to its narrative, algorithms, and images. Its mechanics elicit feelings that complement the narrative without subsuming the differences between our experiences and those of the narrator. At the interface, it holds difference and sameness—and computation and representation—at once.

For Sara Ahmed, affect names the orienting, disorienting, and relational qualities of emotions—the way "emotions work to shape the 'surfaces' of individual and collective bodies" and "how bodies take shape through tending toward objects that are reachable, which are available within the bodily horizon."[31] Following Ahmed and building on the preceding chapter's discussion of Silvan Tomkins's theory of the affects, we can jointly call into question versions of game analysis and affect theory that suggest that analyzing affect and/or video games fundamentally requires a movement away from questions of subjectivity and representation. Rather, how might the "surface effects" of game studies be translated into "surfacing affects"? Returning to the screen as a surface, an interface, a site of touch and being

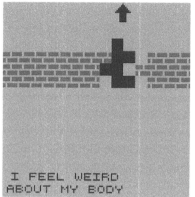 

Anna Anthropy's *Dys4ia* uses the graphics and mechanics of arcade games to give a performative dimension to the feelings expressed by Anthropy.

touched, we might understand, following Ahmed, the ways "the surfaces of bodies 'surface' as an effect of the impressions left by others. . . . In other words, emotions are not 'in' either the individual or the social, but produce the very surfaces and boundaries that allow the individual and the social to be delineated as if they are objects."[32] Affect, like code, is not deep and inaccessible to apprehension; rather, affect and code are two among many processes of inscription—linguistic, technological, and social—and, as such, both are fundamentally imbricated with these other processes.

In both *Dys4ia* and *The Empathy Machine,* this circuit of feeling is not based on some fundamental similarity between bodies and machines but instead is premised on their differences and relationality. When *The Empathy Machine* asks us to touch the screen, k is asking us to feel the difference between a machine simulation and embodied knowledge, between glass and skin. But working through the medium of a video game, she is also using the power of representation through language—programming language and human language—to produce an affective encounter in this moment of touch, of simultaneous connection and disconnection. As the game asks in this awkward moment, might we actually hold "difference and sameness without subsuming either?" *The Empathy Machine* and *Dys-4ia* work as empathy machines not by using technology to create transparent simulations of experience but rather by making us aware of the complicated imbrication of our bodies and our devices, and what we represent through them and about them.

The Interface Is Where It's At

In their account of the complex layering of the biological and the mechanical in Silvan Tomkins's theory of affect, Eve Kosofsky Sedgwick and Adam Frank find its appeal for feminist and queer theory in the interlacing of these systems and the spaces of possibility such human–machine and digital–analog entanglements open up. They write, "Freedom, play, affordance, meaning itself derive from the wealth of mutually nontransparent possibilities for being wrong about an object—and, implicatively, about oneself."[33] In being wrong about where a computer's empathetic potential resides, we might realize the ways that we are wrong about our own assumed innate and "natural" qualities, like our gender or sexuality. Holding a hand against the screen, we might recognize the space for empathy that opens up while we are holding sameness and difference together. This also speaks to the everyday entanglement that is experienced at the level of the intimate interface, where the impasses of connection and disconnection, surface and depth, representation and computation are evoked.

The video game screen as a material surface, a space of representation, and as part of an affective assemblage is a site of everyday intimacy and entanglement. "Entanglement," for Karen Barad, means applying the lessons that the quantum entanglement of matter has to teach us about the fundamental queerness of inter- and intrasubjectivity.[34] She goes deep, to the subatomic level. Tomkins, on the other hand, draws us back to the surface. In what he calls "inverse archaeology," he argues for locating feelings where they appear, on the surface, in the sensory capacities and electrodermal properties of the skin. Barad and Tomkins, like game studies, also share a problem with "representationalism" and want to locate meaning in the matter at hand, not in signification. I use Barad's notion of entanglement and Tomkins's attention to surface in order to argue for a more everyday kind of entanglement, where representation is a form of matter that matters.

The same year that Ivan Sutherland first demonstrated Sketchpad, Tomkins and other psychologists met in Princeton to discuss how the affects might fit into computer simulations of human intelligence. The aim of the conference, sponsored by the Educational Testing Service, was for psychologists to discuss the potential for computers to expand research into the field of personality. In his presentation at the conference, Tomkins used the theories about affect he had been developing to address some of the many problems faced by engineers trying to design computer simulations of human intelligence. The ideal "humanomaton," according to Tomkins, would have an "affect system" in addition to and complementing its cognitive system.[35] In this way he was coyly pushing back against computer science models of "intelligence" that treated it like autonomous information that could be separated from the body. The Princeton conference, like the earlier Macy Conferences, was a node in the emerging network that connected cybernetics and systems theory with the behavioral and social sciences in the mid-twentieth century. As discussed in chapter 1, the cross-pollination among disparate fields at these conferences had profound effects on the expansion of "systems thinking" and computational metaphors into almost all areas of intellectual inquiry in the latter half of the twentieth century. Less well documented and understood was the flow in the other direction—the influence of psychology, the social sciences, and the humanities on computer science and systems theory. As Elizabeth Wilson argues, theories of human personality developed within this cybernetic fold, such as Tomkins's, also influenced the emerging fields of computer science and HCI research.[36] Thus, we might broaden our consideration of the relationship between the ideals of embodiment in HCI and concurrent ideals of bodies as interfaces in Tomkins's work.

What is the relationship between a face and an interface, or between a hand and a digital pointer? About the significance of the face, Tomkins writes, "It is not only a communication center for sending and receiving information of all kinds, but because it is the organ of affect . . . it necessarily is brought under strict social control."[37] In his theory of the affects, the human face is the privileged site of the affects. Tomkins devoted a great deal of his life's work to reading faces for affect and classifying the relationship between what can be read through certain configurations of skin and muscle and what these configurations convey emotionally, physiologically, and socially to our own bodies and to the bodies with which we interact. For Tomkins, the affects are fundamentally generated close to the surface, in the finely attuned muscles and glands of the face. The face and its affective properties communicate simultaneously outward, toward anyone who can see them, and inward, conveying information back into the "deeper" neurological systems.[38] Late in his life, in a public talk, Tomkins explained this further:

> Archaeology digs deep to bring fossils and artifacts from the past to the surface. Our field, which I've dubbed "inverse archaeology,"

assumes . . . that the surface of the skin is where it's at. . . . This conception not only locates motivation on those surfaces where it appears to be, rather than somewhere else which it represents and expresses. . . . It is at once individual and private and social and shared nonverbal communication.[39]

"The skin is where it's at." What Tomkins meant by this is that affect, like code, is often assumed to exist in deep and obscured systems, coming to the surface only as an effect of something set in motion elsewhere. Contrary to this, Tomkins's method, an "inverse archaeology," involved studying the surface, the skin, as a porous interface across which affects are conveyed inter- and intrapersonally.

Notable about this formulation (even if it seems reductive by current scientific understanding) is the way Tomkins uses affect to open up the seemingly closed system of the body, positioning the face as a sensing and sensible surface across which affect flows outward and inward with equal significance, imbricating the biological with the social. He also focuses on the hands—specifically the hands in relation to the face:

The hand acts as if the face is the site of feeling. Thus, when one is tired or sleepy, the hand commonly either nurtures the face, in trying to hold it up to remain awake, or attempts counteractive therapy by rubbing the forehead and eyes as if to wipe away the fatigue or sleepiness. In shame, the hand is often used as an additional screen over the eyes behind which the face may be further hidden from view. In aggression which is turned against the self, the hand may be used to hurt one's own face by slapping it. Children sometimes claw their face with their fingernails. In surprise, the hand is commonly clapped to the cheek or over the forehead. The head or chin or cheek may be scratched when one is uncertain. In great joy, the hand is commonly placed on the forehead or cheek, particularly when there is an element of surprise in the joy. In distress, one hand is frequently placed over the eyes and forehead, and in extreme shock both hands cup the face and hold it up while it is weeping.[40]

In this moving passage, Tomkins describes the way the hands seek out the face as the source of a feeling and work with the face to more fully express that feeling, punishing and disciplining the face or literally touching and holding it in various comforting gestures. While I do not want to collapse the skin into the screen, or vice versa, I do want to say that the touchscreen as interface is where it's at in terms of what is made available to us as *sensible*—and therefore perceptible—beyond the binaries of surface/depth, representation/computation, hidden/revealed. The interface is where *it's* at, if by *it* we mean the everyday intimate encounter where code, images, and subjectivity collide in ordinary but important ways.

To think through the affective encounter of hands on screens, I turn back to *Sword & Sworcery* and its capacity to touch players. In the beginning of the game, the words "Look," "Listen," and "Touch" appear on the screen as vague indicators of how we might sense and make sense of the strange world in which we have landed. The game is equal parts a messianic tale, an homage to *The Legend of Zelda,* and a Jungian dreamscape. From walking through the pixelated fantasy world and interacting with its characters to sword fighting, solving puzzles, and accessing other states of consciousness, we accomplish our actions in *Sword & Sworcery* by tapping, rubbing, and otherwise touching the screen. We play as a Scythian warrior and are told we must complete a "woeful errand" to retrieve the pieces of the Trigon in order to rid the world of an evil spirit known as the Gogolithic Mass. Along the way we are guided by the Megatome, a magic book that allows the Scythian to read minds and traverse the divide between the real world and a dream realm. Holding the game's images, narrative, and mechanical layers together, we see how touching the screen creates an affective assemblage that encompasses us, the game's formal qualities, the code, and the mobile device itself as a layered, penetrable, and more porous object than it often seems.

Sword & Sworcery foregrounds the ways looking can be deceptive and how other sensory capacities, like listening and touching, might augment and expand our knowledge of the world. Looking at the game, the first thing we must contend with is its pixelated imagery. Like many contemporary indie games, *Sword & Sworcery* employs graphics that hark back to the video games of the 1980s. The depictions of landscapes and characters are conscious rejections of the hyperrealistic graphics of most mainstream games. The humans, animals, and landscapes—forests, lakes, mountains, and ruins—all appear to be composed of the hard right angles of low-resolution picture data, like digital images enlarged until their smallest addressable elements, their pixels, are visible. There is much that we can-

In *Sword & Sworcery,* touching the screen creates an affective assemblage encompassing the player, the code, and the mobile device.

not make out in *Sword & Sworcery*'s visuals, including facial expressions, or even faces. The characters are rendered as vague arrangements of pixels. Yet, despite the limited detail, the world of the game is visually quite compelling. The muted forest palette of blues, greens, and grays creates a lush darkness that frames the pinks and lavenders of the moonlit dusk sky.

Listening enhances our looking. Sound effects and the game's music, composed by Jim Guthrie, create an aural landscape that is essential to gameplay. As we learn to recognize Guthrie's musical cues, navigating the world, sensing danger or important clues, and vanquishing foes becomes easier. Listening to the game allows us to perceive things that we cannot make out visually in the low-resolution graphics.

It is the third instruction, touch, that is at the heart of *Sword & Sworcery.* Touch is what allows us to initiate revelatory changes in the images and sounds of the game. Tapping a finger on the screen initiates the Scythian's movement through the world. Tapping again creates more movement. A dark circle appears at the point where the finger meets the screen, briefly inscribing into the game's world the point of contact. Our touch is also made audible. If we touch trees or shrubbery, their pixelated leaves rustle. Tapping the water in the lake disrupts the calm surface. At key moments, we must rhythmically rub the screen to access hidden bridges or other elements needed to complete the Scythian's quest. This updating of the point-and-click mechanic of adventure games uses the affordances of the capacitive touchscreen to render the screen and our manipulation of it as an important narrative element. Following point-and-click conventions, rather than direct avatar control (a seamless relation between the player's input and avatar movement), to move in *Sword & Sworcery* we tap the area of the represented world where we want the Scythian to go. As a convention, indirect avatar manipulation is often understood as a distancing device that pushes the player to identify more with the game world and its objectives than with an avatar. Yet, as discussed at the beginning of this chapter, players of *Sword & Sworcery* report strong empathetic attachments to both the Scythian and the game's entire emotional landscape. This is because the player's touch in the game is not simply a functional mechanic to aid the achievement of the game's objectives. Touch in *Sword & Sworcery* is about movement, but our touch is also moving because it is presented as capable of revealing something deeper behind the screen's and the game's surface.

Across the vast literature on touch, we can identify two major problems and possibilities that thinking about touch makes evident. The first is the diffuseness of touch as a sense. We tend to ascribe it to the sensual capacity of the skin as a sensing organ, but we do not experience touch as limited to the skin. Touch tends to violate any secure notion of inside and outside the body, of giving and taking, of self and other. The second problem and possibility is that touching is fundamentally premised on not touching. For Derrida this is the "law of tact"—that which governs what and who can be touched and how.[41] For Barad, the similar problem of "not touching" is explored through the laws of physics, which dictate that touching—what we feel when we touch something—is actually the force of electrical repulsion, a not touching of matter.[42] Similarly, the capacitive touchscreens of contemporary mobile devices work through touching and not touching. They work through mutual capacitance—that is, the natural electrical charge of the skin, when brought into proximity with the capacitors embedded in a touchscreen, causes a change in the current where the fingers and screen meet. The screen measures and evaluates these contact points to determine the desired action the fingers are communicating. The screen itself is not one solid thing; rather, like skin, it is made up of several layers that interact with each other. The coated glass surface is connected with a bonding layer to two layers of sensing lines, another glass substrate, and then the layers of the liquid crystal display.

Barad builds on this touching/not touching paradox by arguing for an ethical subjectivity and queer politics based on the fundamental dis/continuity and entanglement of matter. She writes, "Hidden behind the discrete and independent objects of the sense world is an entangled realm, in which the simple notion of identity and locality no longer apply."[43] For Barad, the world as it appears, made of discrete objects, is illusory. To counter this illusion, she argues, we must go deep into the subatomic structure of matter to find a new model for empathy based on the fundamental entanglement of matter. In many ways, Barad's model is the opposite of Sedgwick's queer reading of Tomkins. Sedgwick, quite famously, argued for a queer theoretical practice that resists hidden/revealed binaries, or what she called "paranoid reading." She took this lesson in part from Tomkins's insistence that the emotions of the face are not signifying the processes of some deeper system to be *read* on the surface of the body as expressions of that deeper system; rather, they are located, in a very real sense, in the matter of the body's surface. As many do, I find Barad's notion of a "hidden realm" of entanglement that might inform a new ethics very compelling. The modest question I pose, however, is this: Do we have to go that deep—to look beyond the represented world—to feel entangled? In other words, when we tap and rub our fingers across the images of *Sword & Sworcery*, do we need to understand quantum physics to reveal the queer entanglement of the experience, or is entanglement, if you'll excuse the pun, already felt in the matter at hand, the surfacing affects (rather than surface effects) of representation?

The Scythian's first task in the game is to retrieve the Megatome. This book lies deep within a cave that we access through the mouth of a face carved into the side of a mountain. To enter the cave we must rub the screen in just the right way to reveal a hidden bridge, represented as a long tongue protruding from the mountain's face. In this way and others, *Sword & Sworcery* plays narratively and visually with surface/depth distinctions, figuring the movement of the hand on the screen as a diegetic interaction with both a hidden realm within the game's visuals and that which is hidden from us in the game's code—its algorithmic structure. Touching in the right way makes the relationship between the game's code and its images sensible. This is a way of feeling code. It is a kind of intimacy with computational systems that we do not generally experience through other digital interactions. Code is certainly a kind of sorcery if you are not a programmer, or perhaps even if you are. Despite the efforts of critical code studies, most of us interact with all kinds of digital processes without giving much thought to how they work. We usually feel the code only when there is a problem, when it does not work as we expect it to, and the seams of our digital devices and the smooth processes they run are exposed.

The role of the Megatome in the narrative reinforces this representational layering. Functioning as a textual substrate to the game's images (to access the Megatome, we momentarily pause the game), this book of "sworcery" is represented as the code we need to access for clues that will allow us to complete our "woeful errand." The Megatome allows the Scythian to read other characters' thoughts. It is through these characters' thoughts and pronouns that we learn a detail that is not apparent in the game's low-resolution graphics: the Scythian is a young woman. This is not presented as a significant detail in the story, but we experience it as a gentle reminder of the gendered assumptions we bring to video games, our subjectivities as players, and the degree to which the game is addressing these things through a multisensory approach.

As the game progresses we become attuned to the ways we must look, listen, and touch in order to proceed. By testing the reaction of the world to our touch, we learn that the blocky, seemingly low-resolution pixelated imagery also hides within it another visual register that, when revealed, foregrounds a sensual roundness. The Scythian learns to sing a "song of sworcery." In the intimate mechanics of the game, this involves our holding a finger on a radiant halo—called an "areola" in the game—that appears around the Scythian. This touch causes her to enter an altered state of consciousness. The visual effect of the "song of sworcery" is rendered as orgasmic glowing orbs of expanding pink light. In this state we can see and hear things that we could not before and that are crucial to completing the Scythian's "woeful errand." If we continue to apply a successful touch to the game, the Scythian reaches the end, where she learns why her errand is so "woeful." In the game's final battle, no matter how we touch the screen, the Scythian ends up sacrificing herself in order to kill the Gogolithic Mass. Her limp body then floats slowly down the river as people and animals gather along the banks pay their respects.

We do not literally press into the hard surface of the touchscreen, yet in *Sword & Sworcery* we experience of a kind of *pulling in and down*. What we are feeling when we touch the game is, in the most basic sense, code. We are accessing through touch how the game's algorithms work, and we are learning to manipulate the algorithms according to our desire for a certain outcome. But we also feel the game's code touching us back when, in the end, our desire for a certain outcome—the Scythian to live—is trumped by the game's programming, its desire to move us. To be moved by this sad ending, to be touched by the game, is at least partly bound up in the intimate circuit of feeling among player, narrative, and code made possible at the ludic interface. At the screen, we touch the game and the game touches us.

Conclusion

The screen, what it reveals, and what we imagine it to obscure are central problems in game studies, and in the digital imaginary more broadly. As I have explained in this chapter, the ludology versus narratology debate in game studies was not really about the role of narrative in games but rather about the status of the screen and its images and the role of the viewing subject/player in relation to them. As game theorists, we have been compelled to look beyond the screen and its "surface effects" for the "deeper" and supposedly more meaningful structures of code. Similarly, affect theory attuned not to subjects and representations but rather to presymbolic "intensities" has left us vacillating between sites of meaning somehow outside or underneath the symbolic and subjective: the hollow body of "walls" within which code, algorithms, and protocols are now figured as the "deeper" or more significant structures of information connecting bodies and things in networks. Both approaches leave us ill equipped to talk about how representation works affectively across the body, the screen, and the code. Taking "action" as word one for video game theory positions video games as existing ontologically in a relay between the subjectivity of the individual player pushing buttons and the machinic actions and digital processes of the game executing the commands. The screen and all it contains become nearly invisible. Just as "surface effect" analyses of video games ignore the important invisible processes of engagement in play, focusing on code above all else produces a hyperopic vision of digital

culture that literally cannot see the ways digital images tell us important
things about how that code works culturally and about the politics and
aesthetics of contemporary visual culture.

This trend is evident across digital media studies more generally, where
notions of mediation or representation (and the processes of reason and
subjective interpretation they require) are downplayed while the processes of computational information and cognition are emphasized. Questioning this formulation, in this chapter I have returned our attention to the
screen, but through touch, to explore how the screen, as a site of contact
between representation and computation, troubles neat distinctions between these two areas and reveals new possibilities for understanding how
images, interfaces, and code affect us. Touching the screen, a site that usually privileges vision, productively confuses the distinction between subjects and objects. This confusion does not flatten subjects and objects into
the same plane of being but rather taps into (tapping being an important
gesture for the touchscreen) the way we feel and experience this confusion
and intimacy with objects through the kinds of games we touch and the
games that touch us back.

On a number of occasions in *Sword & Sworcery,* when the Scythian is
near the still waters of the lake, the game instructs the player to "Reflect."
The lake reflects the surrounding pixelated scenery, but the smoothness of
the reflection aesthetically references the glossy reflective surface of the
screen itself. The black mirror metaphor discussed at the beginning of this
chapter relies on an image of a black screen—a state that signals that the
screen is not active, maybe even powered off. Yet when we talk about our
engagement with digital devices, our absorption, we are usually not thinking about a black screen but rather about screens that lose a great deal of
their reflective, mirrorlike properties in their function as interfaces for
the media they frame and transmit. In short, our mirrors are rarely black;
rather, they are colorful, full of representations, and covered in our finger-
prints. As such, the black mirror model presents a false distinction between an exterior, subjective, and illusory mode of analysis and a deeper,
obscured, and more powerful and objective mode of analysis. The screen
as black mirror in this model acts as a sharp divider between these two
hermeneutics. Rather, we might figure the screen less as a stable boundary between the computational and representational, the digital and the
analog, or the subject and object and more as a porous "zone of contact"
through which each term is always potentially in contact with and causing

changes in the other. This screen is not actually a firm boundary between two different modes of meaning making; rather, it is a sensual surface that functions within a larger affective system, revealing more than it obscures.

Attention to affect as a method and orientation toward representation can reorient game studies to address the gendered hierarchies and critical impasses that the turn away from representation and subjectivity has reinforced. In turning back to the screen—paying attention to its materiality, its images, and the everyday entanglement it fosters—we can better contend with the unfinished business of representation in theory.

CHAPTER 3

Rhythms of Work and Play

Candies felt like something that everybody would have a positive feeling about. And I wanted something that could have shine and glossiness without being something unattainable.

—Sebastian Knutsson (cocreator of *Candy Crush Saga*),
"Hooked on Candy Crush," Reuters, October 8, 2013

What is eating the zany? Why is she so desperate and stressed out? And why have so many found this mix of desperation and playfulness so aesthetically appealing?

—Sianne Ngai, *Our Aesthetic Categories*, 2012

WHERE DO PLEASURE AND ITS OPPOSITE RESIDE in the colorful and fantastical management of our time? Why does it matter to our aesthetic experience (and I think it does) whether we are matching colorful candies or glittering jewels in a free-to-download mobile phone game? Sebastian Knutsson, cocreator of the massively popular game *Candy Crush Saga* (King, 2012), suggests that we have positive feelings toward candy because it is not an elite commodity. He may be correct. *Candy Crush Saga* has ninety-three million users and makes upward of one million dollars every day.[1] Wherever and whenever people are waiting for something else to happen, *Candy Crush Saga* and other casual games like it are being played: during commutes on buses and subways, in movie lines, in doctors' waiting rooms. A 2012 article in the *New York Times Magazine* concluded that casual games are stupid and are taking over our lives: "Stupid games" the author pronounced, "are designed to push their way through the cracks of other occasions. We play them incidentally, ambivalently, compulsively, and almost accidentally. They're less an activity in our day than a blank space in our day."[2] This representation of casual games as all-consuming but also blank spaces neatly summarizes the way we seem unable or unwilling to attach meaning to them and the time we devote to them. Our play, it seems, is both incidental and fundamental to daily life. Casual games seem too banal *and* too significant to analyze.

I began to think about casual games as a graduate student. While finishing my dissertation, I began structuring my writing time around little breaks and rewards. After I had produced a certain number of words, I allowed myself a brief interlude of nonwriting. Working alone at home in the dead of winter in western New York, I needed the modest rewards I granted myself, although now I find them a little embarrassing to recount here: checking Facebook, having a snack, and playing free casual games like *FarmVille* (Zynga, 2009) on my computer. I was trying to ward off anxiety and paralysis, and the shame and depression that went with them. But I was also trying to find some grace in my work. By constructing a careful balance between work and play, I hoped to find a productive rhythm for surviving the final stretch of graduate school and, more important, some lasting habits for sustaining pleasure in this peculiar type of labor I had chosen for myself.

Classic and contemporary theories of play speak to the balancing act, when we play, between order and chaos and between creativity and destruction.[3] Similar rhythmic patterns are found in many casual games. For example, in *Candy Crush Saga* we work to sort candy by type to score points and progress through the levels. Like all tile-matching games, *Candy Crush Saga* offers pleasure derived from the simple mechanic of creating an orderly set of objects, which, once ordered, disappear, revealing a new disorderly scenario as the setting for the next round. A field of disorder is presented to the player, the player creates temporary order, and a new mess is then presented for ordering. Repeat. In *Plants vs. Zombies 2* (Electronic Arts, 2013) the logic of disorder/order is reversed. The player begins with a clean, empty grid that she fills with strategically positioned plants in neat rows, but then the zombies wreak havoc on the order created. Chaos ensues and, eventually, the player is presented with a clean slate to mess up all over again. Similarly, much of the labor of writing is about creating order from disorder (or sometimes the other way around), working through states of flow and interruption, and wondering if one's revisions are making things better or worse. This experience shaped my interest in casual games and their rhythms of work and play. The affective significance of the games' rhythms, however, came into focus later, when I became interested in affect theory. Understanding the games as rhythmic interludes and then connecting this to the everyday affect that circulates around work, leisure, and the often difficult to articulate longing for a different relationship to work was my starting place for this book.

Casual games speak explicitly to labor and efficiency and to contemporary rhythms of work and play. The small games we play in between other tasks have particular rhythms and temporalities that are bound up with the blurring distinction between work and play in contemporary culture. Casual games offer players more than just work disguised as play; they also offer narratives and rhythms that sensually address the zany conditions of digital labor in the twenty-first century. About zaniness, Sianne Ngai writes, "Like a round of Frogger, Kaboom! or Pressure Cooker, early Atari 2600 video games in which avatars have to dodge oncoming cars, catch falling bombs, and meet incoming hamburger orders at increasing speeds, . . . zaniness is essentially the experience of an agent confronted by—and endangered by—too many things coming at her at once."[4] The word *casual* tends to downplay how zany many of these games are, as if they calm rather than amplify already intense feelings. Affect theory sheds light on a game genre that seems to exist as both ambient media for the individual and a mediating force between individuals and their working conditions and between individuals and their feelings about their labor. In this way, the rhythms of casual games—a quality that exceeds their narrative and mechanical processes—are experienced not simply as emotional states tied to individuals but also as part of the broader affective system that these games make evident and in which they participate. The next chapter considers the capacities of video games to wrest us from our everyday rhythms, to make our bodies move in ways that confound efficiency and productivity, a kind of radical arrhythmia. To get there, we first need to consider the more mundane experience of video games: the casual game as rhythmic interlude between other seemingly more significant activities. Rather than being blank spaces in our day, casual games are affective systems that mediate relations—and our feelings about these relations—between us and our devices, between workers and machines, and between images and code.

The first part of this chapter considers the structural similarities between mobile game applications and productivity apps. The mobility of casual games gives us access to the ways the laboring subject is increasingly tied to the capture, measurement, and commodification of affect. The second part of the chapter looks at the classic time management game

Diner Dash (PlayFirst, 2004) and its numerous sequels and imitators to reveal how these games address working women in particular, and how we can understand their appeal as part of a shared longing for a different relationship to labor in the twenty-first century. Games like *Diner Dash* put the player into an affectively charged relationship to both the working woman represented on the screen and the working body of the player, the work of hardware, and the actions of code. Thinking about casual games as affective systems expands the homologies among the actions of a player's body, the actions of a game's mechanics, and the actions of ideological signification. Affect gets at how casual games *work*—in the sense of the work of bodies, machines, and digital processes, but also in the sense of how these types of games work culturally and ideologically, and how they work on us and work us over in terms of impinging on our feelings, our identities, and our everyday lives.

Video Games and Labor

The industry classification of casual games encompasses several genres—online puzzle, word, and card games such as *Candy Crush Saga, Angry Birds,* and solitaire; simulation, time management, and social games such as *Words with Friends* and *FarmVille*; and less definable hits like *Kim Kardashian: Hollywood* and *Clash of Clans.* These very different games share some basic similarities: they have simple graphics and mechanics, they are usually browser or app based, and they are free or cost very little to play. Perhaps more than anything, casual games are designed to be played in short bursts of five to ten minutes and then set aside. With the advent of micro-purchases, often these games have built-in features that limit the amount of time a person can play in one sitting before being prompted to take a break or pay to continue playing. These games are designed to be interruptible because they are understood to be played in the context of work done while sitting in front of a computer or played on a mobile phone that might at any moment receive an e-mail, text, or call.[5]

The most common stereotype about casual games is that they are played during stolen moments, as breaks or distractions from the paid labor that players actually should be doing. As a genre, casual games bring to mind the bored office worker sitting in front of her computer with a game of solitaire always in progress in the background of her desktop, behind the windows of "real" work for which she is being paid. As such, casual games are affectively charged as guilty pleasures. Some of the earliest games for personal computers, for example, came with a "boss key," which, when activated, masked the current game on the screen behind fake spreadsheets designed to give the impression that work, rather then play, was being done on the computer. The relationship between casual games and the white-collar work environment is widely acknowledged by the game industry. A major industry study, for example, found that one-quarter of "white-collar" workers play video games at work.[6] Given the stigma attached to this behavior, we can assume that the actual number is significantly higher than this self-reported survey indicates. Casual games, by their very definition, are bound up with the temporality of white- and pink-collar labor. Along these lines, Michel de Certeau's example of *la perruque* offers us a way to think about playing casual games as a tactical response to conditions of labor. Literally meaning "wig," *la perruque* is the worker's own work disguised as work for his or her employer. De Certeau writes, "*La perruque* may be as simple as a secretary's writing a love letter on 'company time' or as complex as a cabinetmaker's 'borrowing' a lathe to make a piece of furniture for his living room."[7] Despite his problematically gendered example of the distinction between the simple and the complex, de Certeau understands both such practices as antagonistic to capitalism's uses of workers, and as a strategy through which workers preserve a portion of their labor value for themselves. It is tempting to see casual games in this way—as separate from the work we do for our employers—but casual games are entirely embedded in work culture and rhythms.

Video games have always been an interface between work and play. At their origins, video games were part of the transformation of work in the postwar United States that was heralded by the importance of computers for the military–industrial complex. As discussed in previous chapters, both *Tennis for Two* and *Spacewar!* were created not as distractions from work but as playful means to visualize and make accessible the work of computers. These two games were designed as demonstrations for lab visitors, as ways to enable nonspecialists to see the processing labor of otherwise boring and inert machines.

There is a tendency in analyses of video games and labor to artificially separate the act of playing from the labor that produces the game and the device on which it is played. Play is figured as immaterial and an illusion that shields us from the knowledge of the material labor that went into producing the game itself. In this formulation, video game play is seen as a form of leisure that gets co-opted by neoliberalism.[8] This version of video game history suppresses just how central computer games were to the digital transformation of work from the 1960s onward. Despite the amount of energy that has gone into characterizing the creation of early computer games as countercultural hacking, the games were generally created as part of the broader labor of graduate students and research scientists affiliated with labs partially or wholly funded by national defense contracts.[9] We can see early video games not just as ways to make the labor of computers visible, friendly, and accessible but also—at least in retrospect—as ways of demonstrating and justifying the work of federally funded researchers. As Fred Turner points out, the contemporary hacker/video-game-player-as-creative-worker is an ideal that emerged *from,* not against, the labor contexts of the 1960s and 1970s.[10] This is part of the constellation of labor processes and contexts often called "immaterial labor," which includes the reassertion of the subject and his or her communicative and affective qualities into the worker, self-management, increased precarity, and the universalization of affective labor. According to Maurizio Lazzarato, "What modern management techniques are looking for is for 'the worker's soul to become part of the factory.'"[11] Building on this, Michael Hardt writes, "Where the production of soul is concerned, . . . we should no longer look to the soil and organic development, nor to the factory and mechanical development, but rather to today's dominant economic forms, that is, to production defined by a combination of cybernetics and affect."[12]

Recently, with the gamification of education, work, public health, and other areas of everyday life once hostile to video games, the stereotype of casual games as distracting from productivity has receded. Whether played surreptitiously at work, as part of official job training, or on one's "own time"—say, on the commute between work and home—casual games are intrinsically about the organization, rhythm, habits, and management of time devoted to labor. While "free" time (time spent off the employer's clock) has never really been free, the degree to which our sociality, our bodies, our creativity, and our time are currently harnessed to digital networks that turn our play—rendered as quantifiable and valuable data—into productivity marks a different affective relationship to work and play. What is missing from the immaterial labor critique of video games as well as from arguments for gamification, then, is the crucial understanding of how video games, from the very beginning, were not separate from work but rather a platform for the reconceptualization of work that emerged just as the context of labor in the West was also shifting from manufacturing to information and service industries.

Mobile phones, and their transformation from telephony devices to "smart" platforms for a variety of work and entertainment applications, are also rich sites for understanding the contemporary interrelationship of labor and leisure. Mobile phones are direct technological and ideological facilitators of the diffuse and flexible workforce and are significantly related to the endless workday and "24/7" productive time that began to emerge in the 1970s with ideas like "flextime" and "telecommuting."[13] Melissa Gregg describes the "tyranny of the mobile phone" in her account of the ways mobile digital devices raise expectations for perpetual productivity, regardless of where we are or what time it is.[14] For example, marketers offer us productivity apps for our smartphones that will track and measure the time we spend doing various tasks. Similarly, there are services that measure our social media clout by using algorithms that analyze our activities in online social networks; our scores are then provided to potential employers interested in the value of our networked communication.[15] Apps like Twitter, Facebook, and Instagram encourage us to update our status, post content on the go, and participate in what Sarah Banet-Weiser critiques as the logic of self-branding.[16] Finally, our digital labor is made visible and valuable through the large data mines to which we constantly contribute as we interact with mobile media. Mark Andrejevic observes, "Media and cultural studies, long engaged in the study of media audiences, have tended to focus on new manifestations of audience productivity rather than how these audiences are themselves *put to work* by these proliferating forms of audience monitoring."[17] In recent years, games for mobile devices have transformed the global video game industry and changed how we interact with our phones. Games are the most popular types of apps sold globally. Mobile game apps constitute 63 percent of iOS App Store revenue and 92 percent of the revenue for Google Play.[18] The same devices that have played such a large role in restructuring the space and time of labor are now also often the first devices we consult for leisure activities. Making ourselves visible and making our labor visible in the network become one and the same.

Through mobile phones, the measurement of productivity is outsourced onto the self. Our daily engagement with mobile media, especially games, is a particularly clear example of the productive harnessing of cybernetics and affect. With casual games, our leisure activities take on the representations and rhythms of productivity. In many games, for example, grids organize the action on the screen. In *Plants vs. Zombies 2,* a tower defense game that pits the player's strategic gardening skills against the chaos caused by zombies, the playing grid is a checkerboard of alternating light and dark colors, referencing the game's chess roots. Likewise, in *Candy Crush Saga,* a tile-matching puzzle game set in a Candy Land–inspired world, the candy is displayed in neat rows, with each piece in its own square. The grid, of course, is also a familiar and ubiquitous interface for the organization and measurement of data in many productivity apps. For example, users of apps such as Shift Worker arrange colorful icons on the basic grid of a calendar to keep track of their changing work schedules. Other productivity apps, such as Procraster and Life Graphy, use grids, graphs, gauges, and colorful icons to track and display statistics like number of "productive minutes." Even the names of many productivity apps point to their organizational and self-management rhetoric: OmniFocus, 24me, Grid.[19] In casual games, the grids on-screen are visual links to the rhythms and mechanics of the games' algorithmic structures.

Unlike most other mobile apps, games visually obscure our phones' measurement data and replace them with visualizations of measurement that are specific to gameplay. The various timers, meters, gauges, and score

The organizational grid in *Plants vs. Zombies 2.*

Like many productivity apps, Shift Worker, designed for organizing work schedules, features the grid of a calendar and colorful gamelike icons.

displays in mobile games mimic, compositionally and graphically, the displays on our phones that indicate various modes of connectivity to the network, time, date, and battery power. When we play games on our mobile phones, we seem to momentarily leave the realm of self-measurement and management of productive labor in order to play with and among heightened and fantastical versions of these very categories. In *Candy Crush Saga*, the phone's conventional symbols of connectivity—network status, battery power, and time—are replaced with colorful gauges measuring the player's score, remaining moves, and boosters. In *Plants vs. Zombies 2,* the cartoonish interface measures daylight, plants available, coins earned, and zombies killed. In both games, these visualizations of game productivity are also accompanied by constant prompts to publicize one's accomplishments in the game through social media networks.

The contemporary state of the free-to-play mobile game industry illustrates both how the concrete labor of game production is profoundly alienated from the games' value and our consumption and, simultaneously, how the games that are produced reflect and refract these conditions in quite material ways. Games that cost nothing to download certainly obscure the labor that mined the columbite-tantalite, the assembly of phones under exploitative conditions, and even the labor of game design and coding. Yet labor shows up in the games' narratives, interfaces, mechanics, and algorithms. Through mobile game apps we are invited to engage with systems of measurement and evaluation that produce us not as concrete workers but rather as subjects of the mobile, liminal, and affective temporalities of labor in the twenty-first century. Given the incredible popularity of these games, we can assume that there is pleasure in the ways they register and mediate the liminal time and place of their play. While *Candy Crush Saga* and *Plants vs. Zombies 2* give us access to the curious temporalities of twenty-first-century work and play, they do not quite reveal the particular affective appeal of casual games. What we need to ask about casual games, then, is how do these types of games become aesthetic expressions of increasingly blurred temporalities?

Feeling Casual

Games like *Candy Crush Saga* and *Kim Kardashian: Hollywood* have become ubiquitous and astonishingly lucrative for their creators. Free-to-play, or "freemium," pricing strategies for games have proven to be extremely successful. This pricing model allows players to download games to their mobile devices for free, so they have little concern about whether or not they will like the games. Players may then choose to enhance their game experience by purchasing virtual goods throughout games that have been designed to be endless in order to maximize profitability. Recently, time management games have been folded into the industry classification

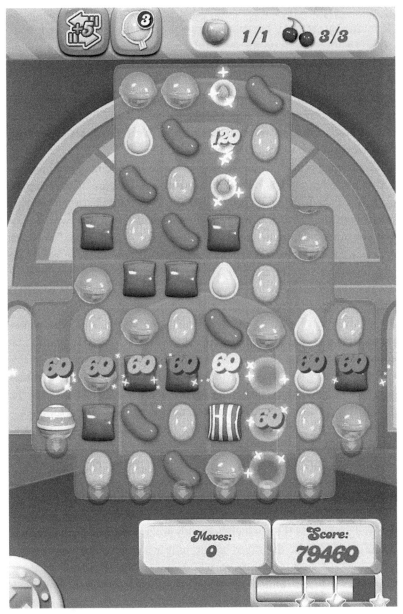

In *Candy Crush Saga* the various icons of gameplay measurement obscure the mobile phone's ordinary measurement icons.

of "invest/express" social networking games, where, Shira Chess argues, a "feminine leisure style" has spread across the casual games market. Invest/express games "count on the interstitial moments that we carry throughout the day, arguing that the player should use those moments deliberately as a mode of leisure."[20] The fracturing of the genre "casual games" into the myriad names now used—time management, resource management, freemium, invest/express, and others—emphasizes a kind of crass instrumentality. These games, according to the names we call them, are about both what we do in them and what they intend to do to us.

Casual games have become one of the most important global business and game design models in the industry, precisely because they reach players beyond the usual video game demographic of men between the ages of eighteen and thirty-five. Many of these players are women who do not play any other video games and would never identify as "gamers." In North America, for example, casual games are the only type of video games for which women over the age of thirty-five have constituted the majority of the market for many years.[21] The casual game market in India and East Asia has exploded in recent years as this affordable gaming model has met the needs of youth seeking games for their mobile devices. Casual games are appealing to an industry that appears to have saturated its traditional demographic. While the already massive budgets for producing mainstream console games continue to expand, the audience for those games has plateaued. Relatively cheap to produce, mobile casual games have the potential to tap into a much wider audience: all owners of smartphones.

While casual games are big business, our cultural understanding and assessment of them remain underdeveloped. Despite their rich and affectively complicated relationship to gender and contemporary labor, attaching meaning and significance to casual games can seem like a frivolous activity. This dismissive attitude is to some extent attributable to the kinds of feelings—guilt, stress, shame, and boredom—that circulate around these types of games and their association with procrastination. But, of course, casual games are also dismissed as culturally insignificant because they are so strongly associated with women.[22] Although the popularity of these games has reduced some of the stigma associated with them over time, they are still held apart critically and formally from other types of video games. Even the term *casual game* itself performs this distinction, ascribing to other types of games, implicitly and by contrast, an aesthetic, narrative, and procedural formalism. With some notable exceptions, casual games are often figured this way in both popular and academic accounts.[23] Theories of games that focus on the actions of code, algorithms, and hardware before narrative and representation are important and compelling developments in game studies. But they are also shaped by a rather narrow concept of what counts as a video game. Such theories seem like overkill when applied to games that require no devoted gaming systems and no honed gaming skills and that, as procedural objects, can appear quite repetitive and obvious. Like casual games, the field of game studies is meaningfully gendered. The divide between representation and computation in game studies mirrors other gendered binaries, such as nature/culture, emotion/logic, passive/active, and humanities/hard sciences, that make it difficult to ask of casual games the questions they demand. Casual games require us to analyze how computation, representation, and the context of play work together to convey cultural meaning.

In *Diner Dash*, the player leads the protagonist, Flo, through a series of increasingly hectic levels as she works as a harried waitress/restaurant owner, seating guests, taking orders, serving food, and clearing tables. One of the top-selling multiplatform downloadable games of all time, *Diner Dash* has spawned numerous sequels and inspired countless imitators, and in 2014 yet another version of the game was released for mobile devices. As such, *Diner Dash* is one of the defining titles of casual games and the time management subgenre. What exactly is the appeal of playing a game about a woman's desperate attempt to please her picky customers with an emphasis on timed tasks and the stressful feeling associated with the piling on of more and more work with each level? Or, to paraphrase this chapter's epigraph from Sianne Ngai, what is eating us, the casual game players, tapping at our screens on the subway? "Why have so many found this mix of desperation and playfulness [the zany] so aesthetically appealing?" Looking at *Diner Dash* we can see how the labor contexts of casual games are explicitly expressed in the narratives, mechanics, and rhythms of time management games. *Diner Dash* and games like it address *and potentially redress* the affective charges around the conditions of digital labor and leisure in the twenty-first century and their gendered dimensions. Rather than seeing these games as blank spaces between more significantly occupied time, we might think about how casual games provide an affectively charged rhythm for the contemporary blurring of work and play.

While it is true that not all casual games are explicitly gendered, and that player demographics are shifting across all game categories, it is also true that the cultural meanings generated through and around casual games cannot be completely divorced from the genre's past and continued associations with women. The extent to which the genre of casual games is perceived as in need of being rescued from feminized mass culture or preserved as a site where women are actually playing video games is less important than the fact that game studies tends to dismiss the entire genre because these seemingly simple games do not fit neatly into an emerging field that privileges procedural complexity, expensive hardware, and graphic realism.

Game theorist and designer Ian Bogost offers a brief but useful discussion of casual games that illustrates the gendered dimensions of taste and distinction in game studies. Asking if there is such a thing as "kitsch" in the video game world, Bogost concludes that casual games fit this description. Using the work of the hugely popular yet critically derided painter Thomas Kinkade as an example, Bogost defines kitsch as "an art urging overt sentimentality, focused on the overt application of convention, without particular originality."[24] For Bogost, kitsch functions as a relationship between the aesthetics of sentimentalism and their display as markers of class location and aspirations of class mobility. *Diner Dash* is kitsch not because it deploys the "naturalistic sentimentalism" of a Kinkade painting but rather because it deploys "occupational sentimentalism" in its depiction of the virtue of hard work. In this way, Bogost sees *Diner Dash* as the equivalent of a motivational poster validating the Protestant work ethic hung in an office cubicle. You cannot hang a video game on a wall, Bogost notes, but casual games like *Diner Dash* are displayed all over the virtual walls of Facebook, publicly marking the players' progress and rewards. He writes, "By surrounding oneself with posters, or games, that espouse ideals of control, the timeworn hope of pure will breeds the wistfulness that makes kitsch appealing."[25] Bogost's dismissal of casual games as kitsch reproduces in game studies the familiar distinctions between "good" and "bad" objects premised on dubious and gendered notions of serious aesthetic value versus sentimentalism. This view of casual games can be understood as continuing the long tradition of dismissing as insignificant cultural forms that are coded as feminine.[26] By comparing casual games to motivational posters and aligning the form with "occupational sentimentalism," Bogost, perhaps unconsciously, points to precisely what is significant about casual games: they affectively appeal to our conditions of labor. Time management games stage the affective labor of being a worker (what it feels like) as well as the work of being a subject who longs to feel differently in relation to work.

Flo(w) and Interruption

Diner Dash's introductory manga-style sequence begins with the text "Somewhere in a dreary office." We see Flo sitting at her desk quietly simmering as faceless coworkers shove more and more paperwork onto her desk. Erupting with frustration, Flo runs screaming past cubicles and out onto the street. Exhausted, panting, and leaning against a building, Flo exclaims, "Man! There's GOT to be something better than THIS!!" At this point she notices a run-down diner that is for sale and decides to quit her stressful office job and open her own restaurant.

Through *Diner Dash's* opening sequence we can see how the narrative and tone represent the working woman and dissatisfaction with the "dreary office." The restaurant, although boarded up and shabby, is rendered in bright primary colors, while Flo's office is depicted in drab grays and browns. Owning a restaurant is figured as a literal escape from the grim cubicles and piles of paperwork. Flo's escape from the office mirrors our own presumed escape from a similar type of dreary work and into the colorful world of the game. The game reinforces this effect on a software level. When *Diner Dash* loads, it automatically takes over the screen, completely obscuring any nongame digital processes for which the device might be used. The entire screen becomes occupied by play. The game's images appear in order to hide our work from view.

Yet, although *Diner Dash* may set itself up as an escape from work, it is also at the same time clearly about working. After we click "Let's Play," the game leads us through a tutorial level where, as Flo, we learn the ropes of being a restaurant owner, which, in the perverse logic of the game, means learning how to be a waitress. The tutorial level also serves as an introduction to the mechanics of the game: the clicking, dragging, and clicking again to achieve the stated goals of quick, efficient, and friendly service. As customers arrive in the restaurant, we must click and drag to move them to tables. Then we must guide Flo through a series of actions, also achieved through simple clicks: taking the order, posting the order for the chef, serving the food, delivering the check, and clearing the dishes. Each successful action earns us points, and we earn bonus points by performing actions in a chain to increase efficiency. *Diner Dash*, like all

The opening of *Diner Dash* both represents and addresses a worker in a "dreary office."

Diner Dash creates a parallel between Flo's escape from her stressful job and the player's presumed escape into the game.

time management games, structures our play through a series of simple repetitive actions that must be completed quickly over timed intervals. The pacing of the game increases and the tasks become more difficult as we progress. As Flo earns more money, she opens new restaurants, each one another step up from the diner.

Like *Diner Dash*, many time management games are explicitly aimed at the working woman and tap into a perceived shared longing for a better working life. Many of these games explicitly simulate occupations through their narratives and mechanics, and many feature working women as their main protagonists. The *Dash* series and its spin-offs include *Hotel Dash*, *Garden Dash*, *Cooking Dash*, *Wedding Dash*, *Dairy Dash*, *Diaper Dash*, *Pet Shop Hop*, *Dress Shop Hop*, *Teddy Factory*, *Betty's Beer Bar*, *Nanny Mania*, *Dr. Daisy Pet Vet*, *Magic Farm*, *Airport Mania*, *Sally's Spa*, *Ranch Rush*, *Hospital Hustle*, *Wendy's Wellness*, and even *Grave Mania* (in which the player takes the role of a zombie undertaker). As the titles indicate, the *Dash* games tend to focus on careers, activities, and interests that are coded as feminine. More often than not, these occupations are portrayed through white female protagonists. The narratives and graphics also tend to frame these occupations as "dream" jobs that the protagonists have come to after escaping less fulfilling jobs elsewhere. Perhaps more than anything, the titles of the *Dash* games speak to the games' time management structures. Each of these games is organized around a mad rush, dash, hustle, or hop to complete repetitive tasks in a limited amount of time. However, as the titles also indicate, playing at these "dream" occupations is not entirely a sentimental endeavor—it is also a mania.

The name of *Diner Dash*'s protagonist also speaks to the perceived goals of time management games and to the flow of efficient labor that Flo (and the player) is meant to embody. In a smart analysis of the *Diner Dash* series, Shira Chess discusses the importance of the time management mechanics for the representation of work in the games. She argues that *Diner Dash* "at its core . . . is a conflation of work and play. . . . While the game is intended for play/leisure time, thematically it involves work spaces that bear a great deal of similarity to work in the non-game world."[27] Chess understands this conflation of work and play as part of the appeal of time management games for women. Citing the work of Arlie Russell Hochschild on working women and time management, she points out that games like *Diner Dash* might appeal to women who already feel the pressure of juggling multiple "shifts" at work and at home.[28] Time management games, according to Chess, are a kind of multitasking for the busy working woman in that they convert leisure time into time management training for their already overextended lives.[29]

Casual game mechanics, narratives, and reward structures seem to speak to a desire for time-bound tasks with identifiable outcomes—things often perceived as missing from contemporary work. There is pleasure in the ways casual games encourage a smooth and efficient engagement with their algorithms. Game scholars often use Mihaly Csikszentmihalyi's concept of "flow," a pleasurable state of absorption in a task, to discuss how video games balance challenges and rewards.[30] Similarly, Braxton Soderman writes: "Csikszentmihalyi envisioned a society that would put these principles to use. . . . When our tasks and activities become more like games, he argued, life itself becomes more pleasurable and also more productive."[31] Soderman goes on to note that employers now look to video games for their models of pleasurable productivity. It is a mistake, however, to think that pleasure in these games (or any games) is derived solely from a feeling of flow. More often, the games are about a frenzied pace and play and/as interruption. What makes a game "casual" is that it functions in the ambiguous times and spaces between the myriad tasks we do on digital devices: between work and domestic obligations, between solitary play and social gaming, and between attention and distraction.

There is a disjuncture between Bogost's description of casual games as kitsch and the analyses offered by Chess and Soderman. Bogost, who is concerned with how the games look, finds sentimentalism about work. Chess and Soderman, who focus on mechanics and the player, find pleasure and social realism in casual games' address to the laboring body. The focus on aesthetics does not address the ways the games' colorful imagery is repeatedly undone by their time management mechanics. And the focus on game mechanics does not get at the odd interplay between the grimly repetitive actions and the cheerful tone. Of course, at the level of image and narrative, there is nothing realistic about the *Diner Dash* games. Beyond the simplistic narratives, the cartoonish graphics, and the uniform whiteness of their protagonists, the games condense the complexity of running a business down to one or two actions. Except for the brief appearance of the chef in the background, we never see Flo's staff. From seating guests to busing tables, the heroic and overworked Flo appears to do it all herself. Knowing this, we can hazard a guess that players are not attracted to these games because they offer realistic representations of their working lives or because they provide simple, sentimental escape. Instead we might think about how the apparent disjuncture between the images and the mechanics is precisely where meaning *and pleasure in that meaning* are produced.

Time management games are also sometimes referred to as "click management" games, connecting the player's manipulation of the interface (clicking a mouse or tapping a touchscreen) with the goals of the game. The player clicks or taps on various tasks to complete them, always juggling multiple tasks and making decisions about order and rhythm in order to complete the tasks effectively. Video game genres are often classified by game mechanics (e.g., first-person shooters, platform games, racing, fighting). As Jesper Juul puts it, game genres are named "after what you *do* as a player, rather than after the fiction."[32] This shores up claims in game studies that game mechanics are more significant to the player's experience than any of the more obvious signifying units. What casual games make clear, however, is that game mechanics, which are themselves kinds of fictions, are intimately tied to the representational practices of games. The actions in a game—what we do—and how we feel about them are shaped by the game's representational fictions, and the actions are themselves signifying practices that create meaning. What can seem like a discontinuity between the banal activity of clicking or tapping on the screen of a digital device and the nightmarish representation of increasingly difficult and endless work is actually a transformation of our relationship to the digital device. The usual search and selection functions associated with the physical act of touching a touchscreen or maneuvering a mouse are replaced with the abstract although quite material repetitive labor of clicking or tapping. The supposedly labor-saving digital device, and the way we feel it and feel about it, is momentarily transformed through play. The rhythmic tapping and clicking we do to complete a timed task is a highly visceral and visible form of work on a smooth machine that is designed to conceal our labor. In this way, our actions in the game make the time and work of the digital device newly legible in ways that reflect how these same aspects of our everyday digital experiences are often submerged beneath the rhetoric of ease, efficiency, and flow. In time management games, where work is both the subject and the presumed context of play, our physical relationship to the machines of our labor is momentarily transformed through this imbrication of narrative and mechanics. Indeed, we experience video games as digital procedures, but our very access to their procedural expression is necessarily couched in and framed by the visual, aural, and narrative dimensions of the games. The opposite is also true. Our experience of a game's representations is always informed by the invisible digital procedures the game asks our bodies to make visible.

The obvious fictions of time management games are about the particular occupations they represent. We play as restaurateur, waitress, farmer, real estate agent, or zombie undertaker, but the actual experience of labor in these games is absurdly easy. The act of harvesting a crop or working an eight-hour shift on our feet is reduced to a series of taps of the touchscreen or clicks of the mouse. On the levels of representation and mechanics, games like those in the *Dash* series can seem like fantasy work spaces. These games portray women as entrepreneurs who are successful because they love the work they do. Their tasks are clearly defined and always rewarded. Their work environments are safe, colorful, and full of zany characters. But the experience of actual labor in the games bears more resemblance to grim Taylorization than to occupational sentimentalism. In these games, the more work we do and the more efficiently we do it, the more complicated, sped up, and vast our tasks become. The games' relative easiness compared to "real" work or hard-core games is deceptive. This is the dash, hop, or mania to which so many of the titles refer. Thus, these games' appeal to players cannot be reduced to a simple notion of pleasure in easy tasks and the satisfaction of achievable goals. At the narrative, mechanical, and procedural levels, *Diner Dash* simultaneously represents a laboring woman, asks the player to perform efficiently on a digital device, and addresses a playing subject who is presumed to desire an escape from the dreary conditions of work.

Casual games' particular combination of flow and interruptibility speaks to the way work on computers and other digital devices is often done. Describing the conditions of digital work, Cathy Davidson writes: "Workflow in the digital age is a constant unsorted bombardment that defies old divisions of labor. We receive urgent memos at a rate never imagined before. . . . And we receive those on the same computer that delivers us banana bread recipes from Aunt Bessie and 'lolcats.'"[33]

Zaniness, Ngai writes, "calls up the character of a worker whose particularity lies paradoxically in the increasingly dedifferentiated nature of his or her labor."[34] Citing the work of Nikolas Rose on neoliberal conditions of labor, she continues:

Post-Fordist zaniness in particular suggests that simply being a
"productive" worker under prevailing conditions—the concomi-
tant casualization and intensification of labor, the creeping extension of the working day, the steady decline in real wages—is to
put oneself into an exhausting and precarious situation. This can
be all the more so in postmodern workplaces where productivity,
efficiency, and contentment are increasingly measured less in terms
of "objective exigencies and characteristics of the labor process
(levels of light, hours of work, and so forth)" than as a factor of
"*subjective attitudes*" about work on the part of workers.[35]

The timing, rhythm, and tone of casual games, from *Candy Crush Saga* to
Diner Dash, interrupt our workflow in precisely the way that interruptibility, fragmentation, and piecework have come to be the common conditions of labor in the digital age. On our computers we move from one
window to another, negotiating the different languages, rules, and logics
of the different software programs we are using. The digital landscape is
not only about the easy flow of seamless touch navigation and information
at our fingertips but also about constant procedural and ergonomic shifts
between windows, programs, devices, interfaces, and lexicons. The everyday experience of digital media is as much an experience of pauses, breaks,
ruptures, and glitches as it is an experience of flow.[36] The digital worker
is constantly asked to move from one task to another, to juggle multiple
and varied tasks simultaneously, *and to feel good about this* as some sort of
improvement over constant focus. Casual games function both as rhythmic interludes between various activities and as emotional mediators
bridging the gaps, pauses, and glitches that are part of our everyday digital
work lives.

   In light of this, casual games can be productively linked to other types
of mass media geared toward women. In her 1970s study of television soap
operas and women viewers, for example, Tania Modleski argues that the
conditions of reception for soap operas correlate with the rhythms of
women's work in the home. She observes that soap operas' highly fragmented, repetitive, and drawn-out narrative structure, as well as the commercial interruptions and the flow between soaps and other daytime programming units, "reinforces the very principle of interruptibility crucial
to the proper functioning of women in the home."[37] Similarly, we might
think of casual games as punctuating and providing a rhythm and timing

to work—whether in the home, at a workplace, or during the commute between these spaces—mediating shifts between different kinds of tasks, different emotional tones, and different people, as well as between attention and inattention. Since Modleski conducted her study, television soap operas have all but disappeared and video games have become a dominant form of mass media. Casual games are filling in for and significantly revising at least one of the cultural functions once performed by the daytime soap. The interruptibility of casual games, their relative simplicity and short levels, offers the player a type of pleasure that speaks to the way her work time is already structured. The experience of rhythm and flow within the games' fictions and mechanics speaks to the desire for a smoother path across the multiple shifts of the day and even across the troubled contemporary work landscape.

Just as *Diner Dash* clearly acknowledges the perversity of the conflict between its cheerful visual fantasy and its grim mechanics, so we might also acknowledge the possibility that the pleasure found in casual games is not based on any simple notion of escape or distraction, or, on the other hand, of social realism. The zaniness of time management games—a quality that lies in their rhythm and aesthetics—is represented by and also exceeds their narrative and mechanical processes. It is felt not simply as an emotion tied to subjects or digital objects but rather as inexorably bound up with larger forces of affective labor and precarity in the twenty-first century. Casual games are the ideal medium for the casualization of labor. Time management games do not simply offer a representation of work, they also offer digital procedures that impinge on, skew, and intensify feelings about work. Through the interplay of their digital procedures, representational practices, and gameplay actions, these games offer a rhythm that addresses a desire for flow in a digital landscape that is defined more by distraction and interruption. Paying attention to how casual games manage flow and interruption across their narratives, mechanics, and presumed contexts of play reveals how they figure as meaningful and affective interludes in contemporary life.

Feeling Zany

In Marxist theory, affective labor is the labor under capitalism that produces and manages feelings—service with a smile, caring for the sick, the products of the entertainment industry. Feminist analyses of affective labor have connected this to undervalued "women's work" in the family and in service industries, such as caring for children and spouses or working as a flight attendant. In both Marxist and feminist analyses, affective labor functions on the level of the subject as the producer and manager of feelings that smooth over the otherwise brutal and alienating conditions of capitalist labor. Chess links the emotional representations in *Diner Dash* to the ways women are called upon to do emotional labor in the workplace and at home: "If the *Dash* games construct a complicated relationship between work and play—then the games, too, have the potential to become a form of emotional labor. . . . And just as emotional labor takes a toll on many women, so might emotional play."[38] We can see affective labor at work in most time management games, from their predominant focus on service-based occupations to how the player's progress is measured and visualized through feeling-based icons. In *Diner Dash,* for example, customers are pictured with series of red hearts to indicate their moods based on the service they are receiving from Flo. Additionally, affective labor can be seen more broadly across the emotional appeals of other types of casual games. If we fail to complete a level in *Candy Crush Saga,* Tiffi, the young girl who guides us through the candy world, is shown crying, with an icon of a broken heart above her. The representation of emotional labor in casual games is only a trace of the affective processes that get called up into representation. These games mediate affective processes that cannot be pinned to a single subject or representational practice, however, and this potentially opens up a space for interpreting the emotional and relational functions of these games beyond a grim assessment of their toll on players.

While casual games are a continuation of mobile media's metrics of labor, they also skew these systems of quantification through their sped-up, exaggerated, and preposterous tasks. Ngai wonders why the strenuous and laborious performances of zany characters in visual culture make us laugh. Her analysis centers on the zany performance of Lucille Ball as Lucy Ricardo in the iconic 1950s American television series *I Love Lucy.* She asks, "What type of aesthetic subject, with what capacities for feeling, knowing, and acting, does this ludic yet noticeably stressful style address?"[39] In the postwar reconsolidation of gendered labor, Ngai argues, we can understand Ball's brand of humor, her hyperactive and usually failed labor performances, as speaking to women viewers increasingly aware of their own affective labor performances in a variety of paid and unpaid workplaces inside and outside the home.

In *Candy Crush Saga*, Tiffi cries and a heart breaks when we fail to complete a level.

Relatedly, consider the ending of the original *Diner Dash* game. After Flo has completed all the tasks to become the head of a restaurant empire, she is transported above the clouds, where a Hindu goddess challenges her to ten waitressing trials inside one of her own restaurants. To enable Flo to complete the trials, the goddess endows her with two additional arms, allowing her to carry twice the amount she could before. *Diner Dash*'s ending is a gentle critique of it's own "endless work" procedural rhetoric. Ngai writes that zaniness "is really an aesthetic about work—and about a precariousness created specifically by the capitalist organization of work. More specifically . . . zaniness speaks to a politically ambiguous erosion of the distinction between playing and working."[40] After Flo has worked her way up and has built a restaurant empire, her reward is extra appendages with which to more efficiently serve. In this way, the game begins by creating an affective relation between player and game based on reward for the completion of small, simple tasks, but as it develops, this affective relationship is transformed by the game's operational logic that, in order to proceed, the player must, in Lucy Ricardo fashion, take on more and more tasks and accomplish them faster and faster.

Recalling the *I Love Lucy* episode in which Lucy and Ethel desperately try to keep up with a candy factory assembly line, *Candy Crush Saga, Plants vs. Zombies 2,* and *Diner Dash* offer players pleasure but also pathos, derived from the predicament of having their sense of control and mastery tested by ever more difficult tasks, increased targets, and more limiting constraints. In *Plants vs. Zombies 2,* the major constraint is time. In fact, the game's subtitle is "It's About Time"—a humorous reference to both the game's time travel narrative and how long fans of the game's first installment had to wait for the second. "It's about time" is also an accurate description of what the game's graphics, algorithms, and mechanics measure, visualize, and make manifestly felt by the player: time and its passing. The game's humor resides in the premise of a high-stakes battle between two entities not known for their speed: zombies and plants. The tension lies in the slowness of the zombies' trajectory toward the player's house and how long seeds take to regenerate before the player can plant them again. As the player progresses through the levels, the waves of zombies and their strength increase, making her job more difficult. For added pressure, the player can tap the fast-forward button to speed up the entire game. *Candy Crush Saga* has a more leisurely relationship to time. The constraint of most levels is the number of moves, not the amount of time, the player has to complete the goal. Yet just as the player gets comfortable with this rhythm, the game presents a timed level. This shift has a visceral effect. The player's heart rate increases as she speeds up her gestures, struggling to adapt to this new constraint and meet the goal set by the game. Just as Lucy and Ethel become zanier in their behavior as the speed of the candy factory conveyor belt picks up, these games and the player's activities within them become zanier and more stressful as they progress. Almost all video games follow this same trajectory, from easy to more difficult as the player invests more time, but in casual games this convention combines with representations, player demographics, and context of use to form an affective system that speaks to the conditions of modern labor and to an unarticulated longing for something different.

The humor of zaniness, Ngai argues, is linked to notions of the laboring subject, who, contrary to Henri Bergson's inflexible subject, becomes preposterously flexible and adaptable under the demands of capitalism. Along these lines, one could argue that casual games—their interfaces, aesthetics, and mechanics—are purely about creating similarly flexible and adaptable subjects. The zaniness of casual games can seem at first glance like utter commitment to productivity; however, as Ngai points out, zaniness—in all its desperate and frenzied action—is also fundamentally unproductive, even destructive. We laugh as, facing the prospect of being fired from their jobs if they let a single piece of candy go by without being wrapped, Lucy and Ethel stuff candy into their mouths and down the fronts of their aprons in an attempt to hide their failure. The destruction of the commodity for the sake of preserving the social relations of labor is another way of describing the zany aesthetic. While playing casual games is always connected to the means and modes of productivity, it would be foolish not to recognize that it is also about taking some time, even if very brief, away from concrete labor and using our machines toward different ends.

The Interlude

The interstitial qualities of casual games—that they are played in in-between times and spaces and that they constitute affective pathways between ourselves and others—are central to their cultural meanings and functions. Brian Massumi describes affect as a kind of intensity that momentarily interrupts the narrativizing linear processes of subjective perception: "It is like a temporal sink, a hole in time, as we conceive of it and narrativize it. It is not exactly passivity, because it is filled with motion, vibratory motion, resonation. And it is not yet activity, because the motion is not of the kind that can be directed (if only symbolically) toward practical ends in a world of constituted objects and aims (if only on screen)."[41] While affect works in the spaces between representation and computation, between the representation of work and the experience of labor in the games, and between the player and the device, it is not as fugitive a process as Massumi describes. Affect lands—as image, as algorithm, as interface—and becomes present and readable to us as feeling, mood, and emotion. If time management games are zany according to Ngai's formulation, they are also sentimental in that they speak to a longing for a different, less fraught, relationship to labor. This inarticulable yet felt longing for a different relationship to work is the space of possibility that affect theory pries open in a consideration of casual games.

As discussed in the Introduction to this book, I understand the rhythmic interludes of casual games as being explicitly related to Raymond Williams's "structure of feeling." It is difficult to classify and explain these moments (the woman playing *Candy Crush Saga* on the way to work, the graduate student harvesting a few crops in *FarmVille* before returning to her dissertation) because they appear to fall outside the institutions that give value and meaning to our lives. They seem insignificant, or even shameful, in their banality. Yet these moments and the feelings they evoke are perhaps the best signs we have of where desire and longing begin to intersect with a call for new social relations. How might a casual game serve as an affective system between the player and a wider community of players?

The "click" or "tap" is both a break and a connection. Clicking on a link on a website or tapping on Flo in *Diner Dash* causes a rupture between the present state of the digital procedure and its next state. Our clicks punctuate the flow of code, inputting new data to close down one field of action and begin another. But, like punctuation, clicks also create an expressive relationship across modes (work/play), spaces (the place of work/elsewhere), and bodies (the player's body/the computer's body and the wider community of players). Since the emergence of the online social network Facebook, the platform has been used as a means to add a social dimension to casual gaming. The rhythm of clicks in a casual game is extended beyond the game itself into the social network. The player's clicking on images, status updates, and friends' wall posts *about* the game *mirrors* the performativity of clicking in the game. Furthermore, when the game uses Facebook to request action in the game—say, encouraging players to trade items with their friends—the player's click resonates across several social and media registers. It actually performs actions in the game, but it also performs a linking action between the player and friends. It is a trivial link, perhaps, but in light of the larger affective processes that circulate around casual games, it opens the player (and the game) up to wider fields of action.

When a casual game is played as part of a social network, the game application uses the network to link players through their shared investment in the game. Through wall posts, the game application encourages players in the same network to buy, share, or swap game items or to set up restaurants, farms, dress shops, and so on adjacent to each other. Game-related wall postings on Facebook are also means to advertise players' progress in the game and to generate competition between players. In the overlap between the affective processes, proceduralism, and representational practices of casual games and those of Facebook, we can begin to see how casual games can create a form of relation between individual players and a wider community of players.

Like all social games on Facebook, *Diner Dash* uses postings on players' walls to encourage regular play. Once a player agrees to allow the game app access to her profile, *Diner Dash* publicly posts nearly constant updates about the player's progress in the game, also posting on her friends' walls to enlist their support. In this way, the game uses Facebook's tool of the wall post both to keep the player engaged with the game, even during times when she is not actively playing it, and to deepen the player's affective investment in the social network. In addition to the ways the game communicates with individual players through wall posts, *Diner Dash*'s application page uses the Facebook status update function to communicate broadly with the game's users, addressing them as part of a wider community of players.

Shortly after the March 2011 earthquake, subsequent tsunami, and nuclear disaster in Japan, many popular online games, including *Diner Dash* and *FarmVille,* and their players used Facebook as a platform through which to express concern for those affected and to raise money for assistance. For example, a status update on *Diner Dash*'s Facebook page stated, "From everyone here at PlayFirst and Diner Dash, our hearts [are] with all those affected by the Japanese earthquake and tsunami."[42] With the promise that profits would be donated to relief efforts, several popular games encouraged players to buy in-game items, such as a pagoda bridge, an "Earthquake Relief Lettuce" crop, or a daikon radish crop. These sentiments and appeals to charitable gameplay were met with more displays of sentiment (mostly positive) by players using Facebook's "like" button, emoticons of hearts and happy faces, and words of support in their comments.

It is fair to take a critical view of the games' mobilization and expression of concern for the victims of the disasters in Japan. Rather then seeing these actions simply as an example of corporations disingenuously using disaster to increase their bottom lines, however, we might also view them as a completely understandable, even expected, outcome of casual games as affective processes that participate in the commodification of affect that is characteristic of twenty-first-century capitalism. The not-quite-articulated longing for different work and a different relationship to labor can be easily captured in the Facebook conventions of "liking" things, sending hearts and happy faces, sharing links, and otherwise signaling the sentiment of global goodwill through digital commodities. The circulation of sentiment we see in this example illustrates how affect gets converted into the displays of feeling that, as Lauren Berlant asserts, are characteristic of the intimate public sphere.[43] As Berlant points out, the intimate public sphere is practiced and expanded through commodities. Casual games are part of the larger affective economy of the intimate public sphere. Rather than dismissing casual games as kitsch, stupid, or blank spaces in our daily lives, we might view them as important contemporary sites of "the unfinished business of sentimentality in American culture."[44] Looking at mass-mediated women's culture mostly in the form of mid-twentieth-century film and literature, Berlant identifies the "female complaint" genre, noting that films and books in this genre "foreground witnessing and explaining women's disappointment in the tenuous relation of romantic fantasy to lived intimacy."[45] She continues:

> Over more than a century and a half of publication and circulation, the motivating engine of this scene has been the aesthetically expressed desire to be *somebody* in a world where the default is being nobody or, worse, being presumptively *all wrong*: the intimate public legitimates qualities, ways of being, and entire lives that have otherwise been deemed puny or discarded. It creates *situations* where those qualities can appear as luminous.[46]

The casual games I have discussed can be productively added to the female complaint genre, although here the complaint expresses not only women's disappointment over lived intimacy but also a whole range of disappointments, not the least of which is the way work culture and labor conditions in the twenty-first century seem to exacerbate gender inequality while at the same time universalizing women's precarious status as workers to massive segments of the population, regardless of gender. Casual games open up the possibility of affective relations that call into question the myths and failures of the digital workplace, the constantly increasing bleed of work into our private lives, and the role of emotional labor in the twenty-first century.

Conclusion

Casual games function in the times and spaces between the myriad tasks we perform on digital devices: between domestic tasks and social obligations, and between solitary private activities and public/private social networks. Through mobile game apps we can see how the blurred distinction between labor time and leisure time is both concretely and abstractly productive. Globally popular mobile game apps like *Candy Crush Saga, Plants vs. Zombies 2,* and *Diner Dash* remediate the measurement aesthetics and mechanics of self-managed labor. At the same time, their zany narratives and game structures leave room for the pleasure of play. When we open a casual game, we open up an affective process, and regarding casual games in this way allows us to see the relationship between their more visible representational practices and their less visible digital procedures. Casual games are also meaningfully gendered, and this is important to keep in mind as we examine how the discourse around these games has been shaped. *Affect* is not a neutral term; rather, it is always culturally situated in relation to the gendering of the bodies and objects of mass-media culture.

Conceptualizing casual games as affective processes stresses the relationship between games as cybernetic systems and their role in larger interrelational systems of representation, labor, identity, play, and so on. This is an approach to culture, like Williams's, that recognizes how much of the sense we make of the world and our actions in it is not entirely caught up in or articulated by clear-cut ideologies or institutions, or by overt resistance. Affect speaks to the spaces, forces, and moments that fall outside the discursive boundary lines of work, home, and social life—say, the moments in the commute between work and home, when we tap our mobile phone screens, playing games to pass the time. These spaces and moments and what they constitute are hard to articulate or theorize, and yet they form the closest thing we know to "everyday life" and a vernacular digital culture. The way casual games both represent the working woman and connote an activity done in order to escape work compels us to understand them as more culturally significant than they are usually made out to be. Rather than being equivalent to motivational posters—static media that simply adorn and reinforce the status quo of class, gender, and labor conditions—casual games are affective processes with the potential to animate changes in these same conditions. This is not to say that casual games constitute an inherently radical or even progressive media form, but it is to say that they participate in a structure of feeling that is different from that of other types of video games, other media forms, and other digital processes with which we engage.

CHAPTER 4

Games to Fail With

Failures, repeated failures, are finger posts on the road to achievement. One fails forward toward success.

—Charles F. Kettering, *Reader's Digest,* May 1989

You fail it! Your skill is not enough. See you next time. Bye-bye!

—*Blazing Star* (SNK, 1998)

VIDEO GAMES ARE OFTEN QUITE FUNNY. Instead of gracefully leaping over a chasm in *Super Mario Bros.,* we fall off the cliff. Instead of moving effortlessly through a virtual space, our stuttering avatar mirrors the inarticulate movements of our hands as we struggle with the controller. Humor in video games also arises from disorientation. It is easy to get lost in games that are set in vast worlds. When a novice player first attempts navigation in a game such as a first-person shooter from the *Halo* series, where he controls both his avatar and the camera movement, unintentional slapstick emerges through the inevitable disorientation. The avatar runs forward while looking backward or gets repeatedly stuck in place because the player cannot yet coordinate the movements of two analog sticks to create a coherent perspective. Even the simplest and most classic arcade games, such as *Pac-Man, Tetris,* and *Space Invaders,* tend to evoke shrieks of laughter. While it is possible to attribute some humor to the images and sounds of these games, these qualities are not exactly what make us laugh when we play them. We laugh when playing a video game, even when the content is not explicitly funny, because we are presented with repeated scenes of our failures in interactions with computational processes.

In other digital encounters this would simply be frustrating, and sometimes it is frustrating in games, but part of the appeal of video games is that they transport our everyday digital failures into the realm of play. Johan

Huizinga argued that "play" is part of a "loosely connected group of ideas" that includes laughter, folly, wit, jest, joke, and the comic, and he suggested that the relationship of these concepts to each other "must lie in a very deep layer of our mental being."[1] Playing video games means willfully setting oneself up for failure. So much of playing these games is an experience of repeated frustration: not slaying the monster, not saving the princess, falling into molten lava, dying. Why do we bother? One obvious answer is that we hope to improve with each go, and then eventually not fail at all: slay the monster, save the princess, not fall into the lava, live. Part of the pleasure that video games usually afford is the experience of redeeming initial failure with success. The games promise that through repetition we might arrive at graceful mastery. This is also the conventional wisdom about failure that is found in self-help books and in inspirational quotations like the one from American industrialist (and inventor of Freon) Charles F. Kettering that begins this chapter. In the American capitalist imaginary, failure—as an externally measured condition and as a subjective condition associated with fear, shame, and anger—is so unacceptable that it must be enveloped within a larger narrative of progress. "One fails forward toward success." Yet capitalism is dependent on failure and its control. It requires that many of us fail, and it also relies on our putting an optimistic spin on failure or wallowing in self-loathing and shame so as not to experience failure as a compelling reason to revolt. More precisely, capitalism can accommodate a tremendous amount of failure as long as its subjects understand their failures as temporary and as the result of personal shortcomings or bad luck, rather than as a larger systemic necessity shoring up an ideology.

But what of failures that cannot be redeemed by fantasies of success? What if, through video games, the feelings associated with failure were put to different ends? Might that experience have some affective and sociotechnological significance that extends beyond the context of our play? This chapter addresses these questions through the work of contemporary artists using video games as a medium through which to experience various kinds of nonprogressive failure. Pippin Barr, Messhof, and Cory Arcangel explore in different ways the capacity of video games to draw us into fail states and what happens when our expectations of video game success are upended. These artists' games use the aesthetics of failure—pixelated and low-resolution graphics, game mechanics that have difficult or awkward controls, the removal of the fantasies of control and mastery altogether—to disrupt the algorithmic and affective trajectories of winning and losing. These games work at the intersection of failure as a feeling, failure as an aesthetic, and failure as programming. From failed images to failed controls to failing to be playable at all, the games make cases for the uses and politics of failure.

"The present is perceived, first, affectively," writes Lauren Berlant.[2] Berlant's work is useful here because, like Raymond Williams's concept of "structure of feeling," it is concerned with how contemporary media can be read for how they make the present *sensible*—how they index, collect, activate, and give shape to emerging and amorphous feelings about broader social conditions. Following Berlant, I consider the works of Arcangel, Barr, and Messhof through the aesthetics of failure and the affects associated with failure to further understand how video games register and express an emerging sensorium of the present. The link between affect and aesthetics as relational and subjective forces is important. It is through the aesthetics of failure in this chapter's archive that we can pick up on the vibrations and gestures of a more widely distributed sensibility to understand how the capitalist relationship to success and failure in North America is fraying and the "conventions of reciprocity that ground how to live and imagine life are becoming undone in ways that force the gestures of ordinary improvisation within daily life into a greater explicitness affectively and aesthetically."[3]

The Shame of Failure

The centerpiece of Cory Arcangel's 2011 show at the Whitney Museum of American Art was a large installation titled *Various Self Playing Bowling Games*. The work consisted of six bowling video games, ranging from an early eight-bit title to more recent games, their corresponding systems displayed on a table, and large floor-to-ceiling projections of the games in action. Standing in the darkened gallery space, visitors watched as the avatar in each game launched ball after ball directly into the gutter. Arcangel rigged the gaming systems to play without a user and to deliberately and repeatedly throw only gutter balls—six screens, six games, and six avatars in a continuous trajectory of programmed failure, doomed to a process of repetition without difference. The sounds of the games bled together, and

Cory Arcangel's *Various Self Playing Bowling Games* consists of six bowling video games, their corresponding systems displayed on a table, and large floor-to-ceiling projections of the games in action. Copyright Cory Arcangel. Image courtesy of Cory Arcangel and Lisson Gallery.

the disappointed groans from their virtual audiences filled the large room with a cacophonous sound track to accompany the failures depicted on the screens. The avatars looked dejected, shook their heads and covered their faces with their hands, and then did it all over again.

In her review of the Whitney show, Andrea Scott writes, "In their epigrammatic structure, Arcangel's looped one-liners recall Samuel Beckett— 'Try again. Fail again. Fail better'—with a little Borscht Belt thrown in."[4] Yes and no. "Try again. Fail again. Fail better" does indeed describe the procedural rhetoric of video games. But in the case of *Various Self Playing Bowling Games,* there is no possibility to "fail better." Each attempt has been programmed to fail the same way each time, negating the optimism of repetition with a difference. The optimism of failing better is part of what Berlant calls "cruel optimism"—that is, an affective attachment to an idea, especially that of "the good life," despite the ways that attachment "impedes the aims that brought us to it [and is constituted by] the sustaining inclination to return to the scene of the fantasy that enables you to expect that *this* time, nearness to *this* thing will help you or a world become different in just the right way."[5] The gutter ball sequences in *Various Self Playing Bowling Games* are not looped through montage; rather, they are the result of the gaming systems' enactment of each game's gutter ball algorithm. This is returning to the scene of the fantasy of success and discovering the cruel fact of code. Arcangel asks us to get used to failure because, as demanded by the games' programming, not a single one of these virtual bowlers is ever going to succeed. Not this time. Never.

Is there anything more central to the American imaginary than expected, even demanded, success? Success, it seems, is always on the verge of arrival and always possible despite many barriers and signs to the contrary. In this formulation, failure is nothing but a necessary step toward success. We fail forward, we might tell ourselves, but let's be clear: failure usually feels like just the opposite. Failure is about fucking up, and it has long been closely associated with entrepreneurial and governmental breakdowns. In English, the word *failure* has existed since the mid-sixteenth century. By the eighteenth century, it was being used to describe both a state of deficiency and lack in an object, process, or institution and *also* this state in an individual.[6] By the twentieth century, Scott Sandage argues in *Born Losers,* his counterhistory of the American Dream, "failure had become what it remains in the new millennium: the most damning incarnation of the connection between achievement and personal identity. 'I feel like a failure.' The expression comes so naturally that we forget it is a figure of speech: the language of business applied to the soul."[7]

We have lost the distance of metaphor. The language of business has become a personal fact. The statement "I feel like a failure" speaks to the personalization of failure and to the affect most associated with it: shame. In his detailed exploration of the negative affects, Silvan Tomkins argues that shame strikes the deepest because it is experienced as a wound inflicted on "the self by the self."[8] On feeling shame after failure, Tomkins writes:

> Suppose one has struggled long and hard to achieve something and one suffers failure upon failure until finally the moment is reached when the head gives way and falls forward, and, phenomenologically, the self is confronted with humiliating defeat. We would argue that cumulative failure might activate anger or distress or even fear, but that in order to activate shame there must be a continuing

but reduced investment of excitement or enjoyment in the possibility of success. Defeat is most ignominious when one still wishes to
win. *The sting of shame can be removed from any defeat by attenuating
the positive wish.*[9]

I quote Tomkins at length here because in this passage and through his
wider conceptualization of shame, he makes a humane case for recognizing the external factors that elicit this internal critique and the ways shame
is a socially conditioned affect that functions to uphold social norms. He
argues that the shame–humiliation affect kicks in at moments when our interest in something (a desire formulated as a thought) is piqued but previous shame-based social conditioning inhibits our interest and engagement
with the desire. Tomkins sums up the shame response as "I want, but . . . ,"
explaining that "the most general sources of shame are the varieties of barriers to the varieties of objects of excitement or enjoyment, which reduce
positive affect sufficiently to activate shame, but not so completely that the
original object is renounced."[10] *I want* to play *Dark Souls* with my friends,
but I'm afraid they will think I suck at video games. *I want* to speak out
against a sexist assumption made by a colleague, *but* I'm afraid I may be
seen as strident. *I want* to say "I love you," *but* I'm afraid that you don't love
me. The desire is expressed in the first part of the sentence, and the shame-
induced barrier to that desire is expressed in the latter half. From the trivial
to the quite serious in these examples using Tomkins's formulation, we can
see how desire is forestalled and dampened, but not extinguished, by internalized shame. We still want what we *want,* but the *but* of shame interferes.

Tomkins provides us with an understanding of the internal and external dynamic of shame that sheds light on how our optimistic attachments
to withering fantasies are registered affectively. His formulation also emphasizes how the deep cut of shame comes from our psychic transformation of external disciplinary barriers (parental and social) into perceived
internal and quite personal failings. As such, his general formulation of
shame can be read alongside the specific types of affective impasses that
Berlant's cruel optimism describes. Berlant is concerned with how our
desires for and attachments to certain objects and ideas persist despite
the fact that our continued investment in them has deleterious effects on
our well-being. For example, she is interested in why young Americans
raised in the context of rapidly declining prosperity for most still cling
to normative ideas of "the good life" based on outmoded notions of up-

ward mobility and economic safety. Tomkins's formulation of the shame affect provides an understanding, in one sense, of what is so cruel about cruel optimism. Our inability to give up the desire (to win, to succeed, to achieve the good life) is proportionally related to how acutely we experience shame for not achieving the realization of that desire. Shame does not extinguish desire—the desire still smolders low and hot—rather, shame fuels the desire by internalizing and making personal the reasons we cannot fulfill it.

This is certainly a time of epic failures—failed states as well as failures in markets, the environment, health care, regulation, and more. And the details of these failures are often intricately bound up with complex digital processes that tend to obscure responsibility while doing nothing to mitigate the effects of the failures on individuals. Compounding this are the ways technological failures have become a daily fact of life—404 error messages, dropped calls, glitches in streaming media, drowned phones, shattered screens, crashed hard drives, not to mention the frenzied pace of planned obsolescence and the related march of our devices into landfills. In 2008, as the global financial crisis snowballed, the phrase "too big to fail" made a sharp and morally bankrupt distinction between individuals experiencing failure (the souls who could no longer pay their mortgages) and corporations that simply could not be allowed to fail. Even as the world becomes more intricately connected and wealth more concentrated among fewer and fewer people, we hold on to the idea that failing to succeed under these conditions is attributable to what individuals do or do not do. This is cruel optimism and the ideological work of shame. Yet in the past few years, Internet meme culture has seemed to reflect and refract this "language of business applied to the soul" by circulating and celebrating individual failure as a way of pointing to the ubiquity of failures, large and small, in daily life. "Epic fail," or simply "fail," memes proliferated in 2008 and continue in the present. The memes were spawned by lingo coined on sites like 4chan that riff on the fractured English of the late 1990s Japanese video game *Blazing Star*, which taunts players with "You fail it! Your skill is not enough. See you next time. Bye-bye!"

It is fitting that our contemporary take on failure—the language we use to describe it, the "fail"—originates with a video game. As Jesper Juul argues in *The Art of Failure*, every video game is about failure on some level, and as such failure is an essential quality of the medium. Juul links the experience of failure in video games to other types of painful

"Fail" memes proliferated after the 2008 financial crisis.

entertainment—the tragic novel, the horror film—and to accounts of these genres that see them as safe containers for painful feelings. One of the reasons we accept repeated failure in video games, he asserts, is that it allows us to experience the feelings of shame and anger associated with losing but ultimately dismiss these feelings as "just a game."[11] Another way of understanding the cultural ascent of video games over the past forty years is to see them as controlled, small-scale versions of the state of endless competition under which capitalist societies live. In his seminal book *Man, Play, and Games,* Roger Caillois argues that the types of games that are popular in a given culture at a given time tell us something about a perceived lack in that culture. Along these lines, Caillois proposes that in contemporary capitalist societies intensely rule-based competitive games (organized sports, for example) might enjoy popularity because they serve as supplements, offsetting a perceived lack of fairness and regulation in the market. Through sports, we might gain a feeling that the winners deserve to win because the game is fair and highly regulated (and the losers have only themselves to blame).[12] Games have now come to structure everyday life, large and small institutions, and life and death, marking a movement away from games as supplements to games as a central and defining mode under capitalism.

Pippin Barr

Pippin Barr excels at making video games that fall somewhere between pranks and structural experiments. They are easy to play yet impossible to win. They are often quite funny, but they can also be militantly boring and sometimes even enraging. Barr's games are freely available on his website, where he also writes about his design process. Many of his games are quite consciously about frustrating their players, and they are framed as such in Barr's exclamation point–laden and sardonic descriptions. For example, the blurb for *You Say Jump I Say How High* (2012) begins with, "Jump! How High! You be the judge! Enter into the magical world of adjusting physics parameters! Just like a programmer! Grow dark with rage! Feel light with elation! Ride the jagged waves of numbers! And collect some coins and stars while you're at it!"[13] Many of Barr's games play with our expectations of how time should work in a video game, using impossibly short or discouragingly long time frames. For example, in *Durations* (2014) the player chooses among mini-games like "One Second Typing Tutor," "One Month Maze," and "One Decade at the Slot Machine," each of which is designed to play at the pace stated in the title, or, more realistically, until the player quits in boredom or frustration. Additionally, Barr's games tend to speak to feelings around larger social conditions of impossible expectations and failure. As the copy for Barr's *Don't Drown* (2014) puts it, "Are you drowning right now?! Don't do that! Or maybe you could just let yourself go! Slip beneath the waves!"[14]

Barr's games often put us in a state of feeling that we are succumbing to the inevitable, giving up and slipping beneath the waves. They are exercises in failure and meditations on futility. A closer look at his games *Let's Play: Ancient Greek Punishment* (2011) and *Snek* (2013) illuminates how failure can function critically at the intersection of programming, play, and aesthetics. At the beginning of *Let's Play* we are invited to choose among five Greek myths. The instructions are basic, and the tasks are familiar: to play as Sisyphus, "rapidly alternate 'G' and 'H' keys to push the boulder up the hill"; to play as Tantalus, "rapidly press 'G' to take the fruit, rapidly press 'H' to drink the water." We quickly realize that there is no way to complete the tasks successfully; we can only futilely repeat them. The pix-elated fruit always moves just out of reach; the water always drains before we can drink it. Just as the boulder reaches the top of the hill it rolls back down despite our frantic button mashing. The game does not track our progress with a conventional display of our score. Rather, it measures our failures. In Sisyphus, for example, our failed attempts to push the boulder up the hill are counted next to the word "Failures" in large blocky letters. More perversely, our repetition of failure and the game's enumeration of it do not bring about an end to the game. There are no "game over" screens in *Let's Play*. Damning us to the endless repetition of failure, Barr denies us even the small satisfaction of completion or virtual death. The only thing we can do is give up and close the browser window, and this is hardly a satisfying option. The game's blurb taunts, "You can do it Sisyphus! Be the boulder! Keep on rollin'! Don't stop! Never give up! No retreat! No sur-render! No end in sight! Just delicious Greek torment as far as the eye can see and as fast as the fingers can type!"[15]

In 2013, Barr translated his approach to failure to a new platform with *Snek*. Designed for iOS, Apple's mobile operating system, *Snek* is based on the classic 1970s arcade game *Snake*. *Snake* was the first game that, be-ginning in 1998, came preloaded on Nokia mobile phones. *Snek* is Barr's homage to this iconic game, but adapted to the world of Apple's mobile interface: the capacitive touchscreen and its multitouch gestures—the swipe, tap, pinch, and reverse pinch and also the accelerometer and gyro-

The player's failures are quantified in Pippin Barr's *Let's Play: Ancient Greek Punishment.*

scope that recognize user input such as tilting, rotating, and shaking. Both *Snake* and *Snek* are simple but challenging games that require the player to maneuver a "snake" around the screen in pursuit of a piece of food without allowing it to run into its own body. With each piece of food it eats, the snake grows longer, making the game more challenging as the snake takes up more screen space. Barr's *Snek* has three basic play mechanics: tilt, turn, and thrust. In turn mode, using a common gesture of the iOS interface, the player maneuvers the snake by turning the device from portrait to landscape and back again. In tilt mode, the player controls the snake by tilting the screen in the direction she wants it to go. Thrust mode—the most physically performative of the three—requires the player to grip the phone and make thrusting motions in the air, like an exaggerated version of the iOS "shake to undo" feature. The difficulty of *Snek* resides in how Barr has programmed the controls to be unresponsive to gentle manipulation, so that the player must absurdly exaggerate the gestures in a mostly futile attempt to make the snake move in the desired direction.

Both *Snek* and *Let's Play* are about eruptions of jerkiness in places where we generally expect smoothness. In *Let's Play* this is most evident at the level of the image. Barr reverse engineered Flash—a software program that uses vector graphics to animate curves and was developed as an upgrade to the pixelated bitmap graphics of earlier digital design—to create a game with bitmap graphics that look like they were rendered on a Commodore 64 computer. Essentially, Barr inserted jaggedness into a program meant to eliminate it. The pixelated imagery is mirrored by the jerkiness of the gameplay gestures. Each mini-game has two-button controls that require no skill or strategy to engage. Rather, the player is compelled to mash the keyboard in spasmodic bursts, alternating between the "G" and "H" keys to roll the boulder up the hill. This disruption of the smooth is even more pronounced in *Snek,* which uses the basic choreography of iPhone gestures but intentionally programs them to work poorly and ludicrously. Documenting *Snek's* development process on his blog, Barr explains that he set out to make a game that is "super awkward to play and thus the opposite of a 'proper' iPhone game."[16] He succeeded in his goal of making an iOS game app that fails to provide the smoothness and ease that we owners of these devices have come to assume as our right, yet he often found himself pulled toward making a very different type of game. Barr writes:

I'd started with the idea of an awkward iPhone game, and suddenly I was making a procedural music-generation game based on swipe input, with nice calming chiming sounds and smooth gameplay . . . as if the device itself demanded [it]. . . . I pulled out of it again and managed to return to making an awkward game, but it's pretty clear that the iPhone (or perhaps iPhone culture) "wants" a particular kind of approach to game making.[17]

The game instructs us to tilt or thrust "smoothly," but smoothness fails us at every turn. *Snek* is not about a slight tilting of the screen, but rather about turning the screen completely away, so that we can no longer see what is going on in the game. It is not about the smooth swipe but rather about the jerky thrust. In fact, thrusting can be so extreme that there is a very real risk of losing hold of the device and launching it into the air.

In *Cruel Optimism* Berlant argues for reading disturbances in bodily gestures, what she calls "glitches," for how they register affective adaptations through historical moments of loss. Berlant is most concerned with losses around what she calls the fantasy of "the good life," by which she means the fraying of "upward mobility, job security, political and social equality, and lively, durable intimacy."[18] She explains, "I want to show how transactions of the body of the aestheticized or mediated subject absorb, register, re-enact, refigure, and make possible a political understanding of shifts and hiccups in the relations among structural forces that alter a class's *sense* of things, its sensing of things."[19] Through *Let's Play* and *Snek*, Barr asks us to perform these glitches. The games use the aesthetics of failure, from the low-resolution graphics to the unwieldy and slightly embarrassing controls, to playfully disrupt the sensual experience of gameplay and to renegotiate our affective relationship to these devices that we are accustomed to using absent-mindedly. The games disturb our assumptions about how games are supposed to work and how our bodies have been trained to interact with them. As such, these games do not lend themselves to helping us pass the time while, say, commuting on public transit. To really play *Snek*, for example, to inhabit and perform its aesthetics of failure, we must take up space and move our bodies in ways that tend to be viewed as odd, even antisocial. Foregrounding this potentially hilarious effect, the opening screen of *Snek* warns, "Best learned in private. Best played in public."

In that video games emerged alongside and are very much a product of the transformations of labor, politics, and our lingering attachments to

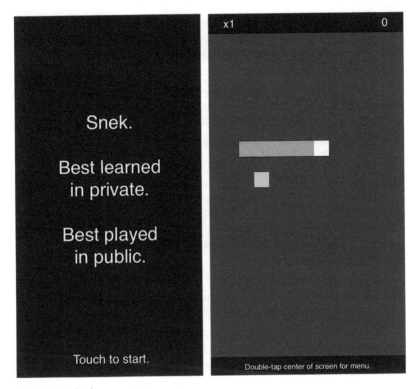

Pippin Barr's *Snek*.

the withering fantasies that Berlant describes, they are excellent sites for detecting and tracing the affective transformations that have accompanied these shifts. Similarly, as Sianne Ngai argues about the relationship between affect and aesthetics, we can understand the aesthetics of failure as emblematic of late capitalism and bound up with everyday mixed feelings about a number of contemporary conditions. She writes:

> These include the implications of the increasingly intimate relation between the autonomous artwork and the form of the commodity; the complex mixture of negative as well as positive affects resulting in the ambivalent nature of many of our aesthetic experiences; . . . the future of the longstanding idea of art as play as opposed to labor in a world where immaterial labor is increasingly aestheticized.[20]

In that Barr's games blur the relation between artwork and commodity and between pleasure and displeasure, they also aesthetically and affectively disrupt the conventional trajectory of failing forward. They ask us to *feel failure differently*—not quite to sit and accept failure, but more to *flail with failure*: to move our bodies nonproductively in relation to machines designed for the more orderly and smooth operations of immaterial labor, in that doing so might reverse the individualization of failure and deflect it back onto the failings of larger systems.

Messhof

Messhof (aka Mark Essen and Kristy Norindr) is a Los Angeles–based independent game design team known for making games with low-resolution graphics that belie their surprisingly difficult controls. Like Barr's, Messhof's games frequently cross the porous boundary between art institutions and the indie game world, and their colorful pixelated graphics hark back to 1980s computer and arcade games. *Pipedreamz* (2010) is all about the fantasy of transforming labor time into leisure time and is also about getting high, surfing, and working at a fast-food restaurant. While most of Messhof's games are distributed for free or for small fees as downloads from their website, *Pipedreamz* is a free browser game. The game was licensed by Adult Swim for its website, which, like the rest of the cable television program's content, is aimed at a young, mostly male audience interested in humor, animation, and popular culture. In *Pipedreamz* we play as an undifferentiated white pixelated humanish figure working at the fast-food restaurant Burgertronix. The game begins with a pink pig wearing sunglasses, who introduces himself: "Name's Steve. I'm your boss, but I'm also yer bro! Meat can change your life. Trust me bro. Lemme see you eat that meat." The Burgertronix tasks alternate between eating meat without being seen and flipping burgers on a grill. Both tasks have fail states. If the other characters see us eating the meat, our avatar vomits a horrifying excretion of pixels and the process begins again. If we flip the burgers too soon or too late, we are taunted by insults from offscreen like "Don't burn them, dum dum."

After we have flipped enough burgers or eaten enough meat, our labor at Burgertronix is rewarded with a mini-game in which our avatar attempts to surf on a huge, seemingly endless blue wave. Strangely, it seems like the meat at Burgertronix has the potential to transport us to an alternate

Pipedreamz sets us up for failure in work and play.

surf-based dimension, where Steve, our pig boss, can also be our "bro." *Pipedreamz* alternates between these scenes of abject work and seemingly idyllic leisure, punishment and reward. Yet *Pipedreamz* sets us up for failure. The work we do at Burgertronix is quite easy to master. The controls are simple and, as in Barr's games, button mashing is about all that is required. Conversely, the leisure section of the game is quite difficult. The surfboard moves too fast across the water and is difficult to control in any skilled fashion. We fall into the water and have to swim to the surface again and again. The alternation between success and failure at first seems to replicate the same basic give-and-take that structures most video games, but in *Pipedreamz* this alternation never resolves itself into anything resembling a winning state. No matter how many points we earn or how well we flip burgers, we always end up getting fired.

The difficulty of Messhof's games is often their most noted feature. Ed Halter compares the use of strange controls and difficult navigation in Messhof's games to "the way a musician might use carefully orchestrated noise."[21] In its refusal of the basic expectations of rhythm and melody in music, noise, like the controls and graphics of Messhof's games, disrupts the clear signal of "winning" and our assumptions about what constitutes the form and our enjoyment of it. This is also true of Messhof's most widely known and critically lauded game, *Nidhogg*. Designed and released by Essen in 2010, *Nidhogg* is an extremely difficult sword-fighting game for PCs, rendered in colorful glitchy graphics. Players compete with each

In *Nidhogg,* the simple graphics belie the game's difficulty.

other, in multiplayer mode, or with the game's AI in single-player mode in a tug-of-war battle that allows the player to progress across the map to the finish line only after repeatedly killing multiple adversaries. The game is difficult to play well because it is hard to control the small, blocky figures. Their undifferentiated forms contradict the speed and precision of their movements.

We feel failure in *Pipedreamz* and *Nidhogg* in their juxtaposition of simple graphics and difficult mechanics. Yet it is not a simple bait and switch. In Messhof's games, the instability of the image—the possibility that it may disintegrate, obscure its dangers, or erupt into a colorful and often comically gruesome display—is an ever-present threat that necessarily informs how we play. In *Nidhogg,* for example, we learn early on that our sword-wielding avatar and opponents, once pierced, will bleed continually accumulating arrangements of pixels that will end up obscuring the pitfalls of the playing field, leading to more death. Similarly, like our bodies come undone, the pixels of the map itself may dissolve underneath us, leading us to fall to our death. In *Pipedreamz,* the pixels are more stable, but their crudeness and uniformity challenge us to decipher the contours of danger. While surfing on the enormous blue wave, we are encouraged to try various tricks for extra points, but the visualization of both the execution and the effects of those tricks is beyond the capacity of the game's low-resolution graphics. We fail at consistently pulling off tricks and we fail at improving because the game's images fail to capture the detail of the game's mechanics. Unlike in Barr's games, in *Pipedreamz* and *Nidhogg* our actions have causes and effects that lead to some sort of end. What comes radically undone in the relationship between the visuals and the mechanics are the expectations of *fairness* and *consistency* that undergird most digital gameplay.

Aesthetics of Failure

Video games are not distractions from the frustration and failure of our everyday lives; rather, they are intimately linked to how we feel failure— not just how we feel *about* failure, but how we actually experience the feelings associated with failing. If the impasse, in Berlant's terms, is that we are caught up in the desire to succeed according to the unrealistic terms of late capitalism but our attachment to that desire is also what keeps us from finding a way out of that impasse, then video games as the ur-medium of failure might provide another scenario and another way of sensing the present. Can video games be a medium for feeling differently about failure in a society that demands success but provides fewer and fewer paths to it?

Tomkins offers a partial, if unsatisfactory, answer when he writes, "The sting [of shame] can be removed from any defeat by attenuating the positive wish."[22] Similarly, in *The Queer Art of Failure*, Jack Halberstam makes a case for embracing failure as a tactic for short-circuiting and undoing the heteronormative benchmarks of success in American culture:

> Failure preserves some of the wondrous anarchy of childhood and disturbs the supposedly clean boundaries between adults and children, winners and losers. And while failure certainly comes accompanied by a host of negative affects, such as disappointment, disillusionment, and despair, it also provides the opportunity to use these negative affects to poke holes in the toxic positivity of contemporary life.[23]

Is the answer to Berlant's formulation of the impasse around cruel optimism to give up our desire for success? The short answer is no. The expansion of our capacity to feel failure differently, to flail with failure, is a project qualitatively and sensually different from embracing failure as radical politics. It is unclear in Halberstam's formulation how failing and its negative affects can meaningfully intervene in neoliberal structures that seem to thrive on failure. As Jamie Peck argues, rather than being derailed by the failures of its internal logic, neoliberalism mutates and fails forward.[24] Furthermore, neoliberal policies combined with technocratic governmentality exacerbate inequalities and position some to fail better and with less risk than others. For most, the endgame of failure, even radical failure, is still internal, personal, and shameful failure. The games I am discussing modify the affective charge around failure and disturb shame's internalized trajectory. They resist immersion and smoothness and instead point to the programming of failure. To flail with failure, as we do in these games, is to spasmodically disrupt the quiescence of the impasse and to become attentive to the relationships among our failings, our feelings, and the systems with which we interact.

Affect and aesthetics are intimately related. The term *aesthetics* is derived from the Greek word *aesthesis,* meaning "sensation" or "feeling." Both affect and aesthetics seek to describe the visceral and immaterial ways we form relationships with and make meaning from our surroundings. Yet you would not know this from game studies' accounts of aesthetics. Although the significance of the ludology versus narratology debates in game studies has been overstated, these debates had lasting effects on how aesthetics are defined and treated in game studies. The term *game aesthetics* is generally used to refer to how games look—the style of animation, the heads-up display, the graphics, and so on. Aesthetics get collapsed into images. Related to the broader division between image and action in game studies discussed in chapter 2, there is a tendency in game studies to cordon off aesthetics from the computational and interactive aspects of games.[25] This rather arbitrary separation aids approaches that wish to emphasize the ludic aspects as the essential feature of video games and connect them to the history of all types of games in order to disentangle video games from other visual media and theories of representation.

The reduction of aesthetics to images and the relegation of images to secondary importance, after action, in game studies makes for a rather anemic application of aesthetic theory to the medium. Graeme Kirkpatrick makes a case for an "aesthetic approach" to game studies that accounts for how our pleasure in games is bound up with the whole experience of what it feels like to play a game, arguing that aesthetic concerns "cannot be bracketed off as incidental to gameplay but must be understood as central to and organizing the whole activity."[26] Yet even this more holistic approach goes off the rails when Kirkpatrick tries to contend with the assumed primacy of action over representation in game studies. He does this through a radical separation of form and content. He writes, "If we want to understand how the video game has come to occupy the position it has in contemporary experience we need to analyze it in terms of its form; the way it shapes space *to create the possibility of meaning only to stop short of actually providing it.*"[27] Formalism allows Kirkpatrick to make a case for the importance of video game aesthetics without having to account for their representations. He sticks to the beauty of high-resolution graphics, the dance-like movements of players' bodies, and the rhythms of algorithms to propose the video game as a container for aesthetic experiences that emerged in the 1960s to fill the void left by conceptualism's abandonment of form for content. Kirkpatrick's account of video game history (and conceptualism) implicitly condemns art for politicizing aesthetics and suggests that video games, by contrast, must occupy an apolitical aesthetic position in culture.[28] Cementing this view, Kirkpatrick writes, "If the video game is such a container [for aesthetic experiences] then the questions raised by its success for cultural politics are not the ones we thought we had to ask when we were preoccupied with whether games are violent, or sexist, or whatever."[29]

What is at stake in this vigorous denial of visual and political meaning in video games? In light of the attacks on feminist critics of the video game industry and culture, and contrary to Kirkpatrick's opinion, it seems increasingly important to understand how video games are more than just containers for nonideological aesthetic experiences. Game studies needs to broaden its understanding of and engagement with aesthetics to take into consideration the ways aesthetics are always a negotiation between that which is seen and that which is felt, and this necessarily has effects on the mediation of identities and subjectivities. Our ability to laugh freely at our video game failures is in some ways constrained by how we already feel positioned by gaming culture's subjects and objects. Video game mastery, what constitutes a game, and who *are* and *are not* video games' proper subjects are often gendered and raced boundary markers. The degree to which one plays the "right" games with the "right" skills leads one—as though following a flowchart with binary yes/no answers—to arrive, or not, at the identity "gamer." To be *too* disoriented in play is to fail the wrong way, to be unfunny, to just be a bad player. As such, feminist approaches to video games are wedged uncomfortably between game studies' problematic turn away from the screen and its images (limiting the relevance of feminist engagements with representation in games) and online gaming culture's decidedly unfunny attacks against women (where being a feminist is to be the worst kind of failure). Hence, the meanings we make of games and their capacity to affect us are part of a complex interaction of factors that exceed both representation and mechanics. Aesthetics are not a particular aspect of video games but rather a way of identifying an affective relationship that is created between and across the games' images, sounds, mechanics, hardware, algorithms, and players. To understand the expressive potential of video games' aesthetics of failure, we cannot separate representation from computation, or the screen from the code.

The interplay between images and mechanics is a core feature of how the games I discuss in this chapter express failure. The images are not secondary in their affective meditations on failure; rather, they are the first point of contact and condition our ambivalent responses. At the level of the image, Barr, Messhof, and Arcangel reject the smoothness and cleanness of contemporary digital design. They very consciously traffic in outmoded and cheap digital aesthetics. Messhof and Barr both use inexpensive, nonprofessional Flash-based game design software that lends their games an amateurish and nostalgic appearance. Arcangel appropriates and modifies video games from his youth and, under the banner of "dirt design," eschews the professionalization of contemporary web design and instead celebrates the vernacular aesthetics of early web culture. Although they do not use the term, all three employ an "appropriate technology" approach that is rare in an era when that kind of thinking has more or less vanished. In a sense, they fail to progress to high-resolution and hyperrealism and remain stubbornly in the computer's past.

The pixelated and glitchy graphics in the games are related to broader gaming trends. Browser games, mobile game apps, free or cheap game-making software, and new funding and distribution opportunities for independent game designers are all part of larger recent shifts in the global video game industry: the shift from games as physical objects to downloadable files, the shift from consoles to mobile platforms, the shift from hard-core to casual gaming. These shifts have attendant aesthetic effects.

Reining in the race to hyperrealism and complexity in the video game industry, smaller platforms and digital distribution tend to require simpler graphics, mechanics, and interfaces.

We can link the work of Barr, Messhof, and Arcangel to a wider movement in contemporary art, christened the New Aesthetic, that embraces errors, glitches, outmoded hardware, old video games, vernacular digital culture, amateur graphic design, and other aspects of digital media that are generally dismissed as obsolete or simply "bad."[30] As demonstrated by works ranging from Takeshi Murata's glitchy videos and Kelly Goeller's pixelated street art to the degraded YouTube and cell phone imagery of Johan Grimonprez's videos and Jon Rafman's Google Street View photography, the New Aesthetic is a response to the overly commercial and derivative design aesthetic of mainstream digital culture. While not all of the work is about failure per se, grouped together the various stylistic approaches of New Aesthetic artists offer ways of seeing and feeling the world that embrace the mistake or the badly rendered in order to reject the hegemony of the smooth digital aesthetic and its links to what Jodi Dean calls "communicative capitalism."[31]

Barr's, Messhof's, and Arcangel's games can be read as rejections of the smooth digital aesthetic. The simple graphics and low-resolution imagery set us up for an easy win, which we are then denied by the difficult mechanics. At the level of gameplay, they use difficult controls or, in the case of Arcangel, remove the player's control altogether. This jaggedness in the image and jerkiness in the controls constitute what I call the aesthetics of failure. The aesthetics of failure speak to the technological constraints of nonprofessional game design software and smaller platforms but also to the ways failure in games can be a preferred aesthetic and an ideological tactic.

Mirroring their jagged and twitchy aesthetics, these games ask us to perform various arrhythmic, jerking, and flailing gestures as we attempt to bend the games' algorithmic logics to something resembling a win. Yet we cannot win. The interplay between representation and computation is essential to how the aesthetics of failure work in these games. They disrupt the commonplace harnessing of our bodies to machines that are meant to smooth and expedite labor. They throw a wrench into the way shifts in capitalist production—from Taylorism to the self-management in immaterial modes of labor under neoliberalism—have depended on a smooth physical and affective relationship between workers and machines.

Programmed to Fail

Recently, video games that experiment with intentional failure have achieved cult success. Games like *Octodad* and *Super Gang Beasts* use the aesthetics of failure for comedic effect. Part of an emerging genre sometimes referred to as "fumblecore," these games are designed to be intentionally and comically difficult.[32] In *Octodad* and its sequel, *Octodad: The Dadliest Catch,* we play as the title character, a cephalopod masquerading (poorly) as a human suburban dad. Our goal is to move Octodad so that he appears as a "normal" man and does not reveal his true nature. But the control scheme is so intentionally complex and sensitive that, despite our best efforts to contort our hands correctly, Octodad tends to flail through his suburban world, smashing his tentacles against everything in his path. The games are funny because they brilliantly foreground the limitations and disorientations of our bodies entangled with the code and hardware of the games.

We tend to laugh at certain types of failure—in games or otherwise—because we get caught up in what Sianne Ngai describes as their "zany" aesthetic. Ngai revisits and revises Henri Bergson's notion of the comic arising from "mechanical inelasticity." According to Bergson, we laugh at the folly of a person who, through a kind of machinelike repetition of behavior, cannot adjust his body to a changed circumstance.[33] Bergson gives the example of a man who stumbles on a rock while walking down the street and falls down. If the man had decided to just sit down in the street, adjusting his body in relation to the street as if he had fallen, we would not laugh. But the appearance of a vital and living being moving in a routinized fashion, walking, and not being able to change this routine to account for an obstacle produces our laughter. We laugh at the zany, Ngai argues, because of *too much* elasticity. Ngai proposes a historical shift from Bergson's time to the present, when an aesthetic of zaniness emerged to comically address the state of trying to wholly adapt our bodies to the machines and technics of capitalism. She asks, "If the rigidity of others is what makes us laugh, can an absolutely elastic subject—one who is nothing but a series of adjustments and adaptations to one situation after another—be genuinely funny?[34] The zany performer evokes the laughter, but also the discomfort, of watching someone too willing to bend and respond to the increasing demands the world makes on her body.

The zany in video games, then, is often about the mechanical limitations of both bodies and machines that surface through their necessary entanglement at the interface a game provides. Hence our laughter at Octodad's rubbery body and our inability to control it, or our laughter as we frantically smash our hands against buttons or screens to respond to the increasing speed and demands of pretty much any game. Video games, which emerged in the mid-twentieth century as affective interfaces for the cybernetic interactivity of computers that was transforming the landscapes of work and leisure, are then where we find the zany in its most advanced state. Humor in video games is not really an effect of our witnessing the "mechanic encrusted upon the living," as Bergson puts it.[35] Rather, more aptly, humor in video games emerges at the moments when the limitations of our living bodies meet the limitations of the software and hardware of a game—the implicit humor in the reductive statement "Press A to jump." This is more akin to humor arising from our witnessing the living attempting to thrust itself upon the mechanic. Plugged into a cybernetic loop with the game, we imagine that we can transfer the full potential of our embodied vitality to the mechanics of the game, but, flailing, we repeatedly fail.

Since he started making art in Buffalo, New York, in the early 1990s, Arcangel has displayed interests that have vacillated between breathtaking virtuosity and abject failure. For him, failure seems to exist in the gap between our ability to use digital tools as amateurs and how quickly digital technologies and aesthetics seem to become outmoded and "bad." Planned obsolescence. Programmed to fail. Devices fail, corporations go out of business and stop manufacturing software, and the trends of Internet use shift, as do their aesthetics, leaving closets and landfills full of "useless" devices and hard drives and energy-hungry servers full of outdated files, software, and websites that have not been updated since the 1990s. Arcangel's work expresses an overt desire for the vernacular digital aesthetics of early web culture, before web design templates, blogs, and social media obviated the need for everyday experimentation. Arcangel's official website communicates a lot about his relationship to web culture. Cory Arcangel's Official Portfolio Website and Portal is intentionally naive and harks back to the days of "home pages" and "web portals."[36] It is cluttered with advertising and looks hopelessly outdated. The frames, fonts, and animated GIFs locate it, designwise, in the mid-1990s. Arcangel uses outmoded web design aesthetics to expose the contours of a form that tries very hard to be transparent.

If the most essential feature of video games is, as many claim, that they create spaces for the players' actions above and beyond their programmed narratives and images, Arcangel's *Various Self Playing Bowling Games* installation casts doubt on this notion. These games do not address us as potentially masterful players; rather, they address us as observers of their systems and ask us to see them as authored and programmable systems. The circuit board is crucial here. The circuitry that allows the games to be "self-playing" is exposed and inscribed by Arcangel. Its crudeness contrasts with the hidden mechanics of the molded plastic consoles and reflects on the "madeness" of both. Arcangel's manipulation of the games and removal of player control bring to the foreground the ways the games' designers have programmed failure to look, feel, and sound. As such, the installation offers us the most radical version of flailing with failure by alienating us from the subject position of player and its attendant fantasies of mastery. As "self-playing," the games fail to offer the subject position of "player" to the viewer. In the gallery we are watching the machine enact the failures of imaginary players. As viewers, we are completely outside the cybernetic loop, a position that violates the most basic (if exaggerated) principle of video games: that we can control them. Denied the possibility of playing the games, we contemplate the games as systems. We notice how with each new generation the player has more vectors of control over the ball's trajectory and measurement of success. The graphics and perspectives of the later generations directly mimic televised bowling competitions, and heads-up displays show interfaces for adjusting the angle and spin of the ball, while graphics convey measurements for "power" and "accuracy." This obsession with remediation and accuracy in the later games makes the repeated gutter balls all the more absurd. Through the repetition of the representation of the avatars' failure to score even a single point, we are provoked to think about the games as states of failure on many levels. Even though the graphics become somewhat more realistic and three-dimensional over the chronological progression, we cannot help but notice how they fail to capture realistically the appearance and movements of human beings. The games also fail as simulations. No matter how realistically they represent the space and sounds of the bowling alley, they are at best remediations of televised bowling rather than simulators of the actual physical experience of throwing a heavy resin ball down a narrow wooden lane toward ten pins.

Repetition is key to Arcangel's work, but it is repetition without differ-

Arcangel's signed circuit board foregrounds the programming of failure. Copyright Cory Arcangel. Image courtesy of Cory Arcangel and Lisson Gallery.

ence. Countering the logic of repetition in video games where we repeat sequences and levels in order to make the outcome different, repetition (of button sequences, programming, and software inputs) in Arcangel's work denies the logic of repetition for the sake of progress. What if we repeated failure, like Arcangel's virtual bowlers, with no measurable movement toward anything resembling success? This is repetition to stay right where we are, refusing originality, refusing the new, and repeating until the accumulation of repetition exposes something about the process itself. By removing the possibility of progress, Arcangel's self-playing games repeat lines of code until, as when we repeat a common word over and over again, its strangeness is revealed.

Conclusion

Play, and its long association with creativity, is an important cultural hinge between aesthetics and affect as subjectivizing forces. Affect and aesthetics both provide ways of delineating the contours of a subject or object in relation to other subjects or objects as well as the many points of overlap that join us emotionally to others. The aesthetics of failure speak to what Ngai describes as the subjectivizing force of aesthetic categories in that they "play to and help complete the formation of a distinctive kind of aesthetic subject, gesturing also to the modes of intersubjectivity that this aesthetic subjectivity implies."[37]

In the early part of the twentieth century, Walter Benjamin argued that the function of film is "to train human beings in the apperceptions and reactions needed to deal with a vast apparatus whose role in their lives is expanding almost daily."[38] Benjamin's concept of *Spielraum,* or room-for-play, in cinema refers to a number of qualities of the medium in which he saw the possibility for radical social effects. He imagined room-for-play in film to change the relationship of humans to machines through a kind of retraining. Miriam Hansen writes, "*Spiel* . . . provides Benjamin with a term . . . that allows him to imagine an alternative mode of aesthetics on par with the modern collective experience, an aesthetics that could counteract, *at the level of sense perception,* the political consequences of the failed—that is, capitalist and imperialist, destructive and self-destructive—reception of technology."[39] This reference to the "failed" reception of technology is a massive understatement considering the alienation, exploitation, and loss of life and limb associated with the new mechanized forms of labor and warfare in Benjamin's time. Hansen links Benjamin's interest in play to his wider project, carried through much of his writing, "to theorize an alternative mode of apperception, assimilation, and agency which would not only be equal to the technologically changed and changing environment, but also open to chance and a different future."[40]

Video games have inherited and substantially revised our perceptions of technology and the meaning of play over the past sixty years. As Peter Krapp notes, some sorts of games generally come preinstalled on our digital devices to "teach us how to handle them."[41] But games, it is important to understand, are also part of a process of training us how to feel about these devices and the larger systems for which they are proxies. If cinema offers room for play, in Benjamin's terms, video games offer that as well, but they also offer room for error and failure.[42] This is not to say that video games function as didactic mechanical interpellators; rather, it is to point out that Benjamin's famous essay can be read beyond the limits of a strict Frankfurt School interpretation for the way it speaks compellingly to the intersecting forces of aesthetics and affect in an emerging medium, and to how these are intimately related to how we feel about the medium and, in turn, how the medium provides a particular emotional texture for its time. Krapp notes that the room for error in games functions like a release valve in the hyperbolic computing discourses of order and infallibility, but it also trains users to assume that any errors are their own in the human–computer interaction formulation of the "user error." The games discussed in this chapter, however, use the aesthetics of failure and their affective dimensions to question the logic of "user error." The intersection of aesthetics and affect in these games gives us access to the ways the senses make meaning through cultural objects, like games, beyond the scope (and the scopic limits) of representation. This intersection also moves us beyond a tendency in game studies to artificially separate aesthetics from meaning making.

The aesthetics of failure in Barr's, Messhof's, and Arcangel's games speak to the sociohistorical context of failure and also to our yearning for different relationships with our machines and their sensual capacities. Their games are asking us not to celebrate failure but to flail with it for a while and learn its contours. Maybe flailing with failure through games that jerk our bodies out of the smooth rhythms of frictionless labor, or deny our participation at all, will shift our attention away from perceived personal failings and back to the failures of a larger ideological formation—say, a user interface, a digital platform, or even an economic system. The aesthetics of failure in these games create different affective conditions and trajectories for failure in a culture that keeps raising the bar of success while simultaneously pulling the rug out from under all but the most privileged.

Conclusion

Affective Archives

"GAMES DON'T MATTER," writes Miguel Sicart. "Like in the old fable, we are the fools looking at the finger when someone points at the moon. Games are the finger; play is the moon."[1] Sicart contends that, contrary to the classic theory of play inherited from Johan Huizinga that separates acts of play from the rest of life, play is what puts us into the world and what permeates all aspects of it. "Play is being in the world, through objects, toward others."[2] This book is one contribution toward furthering our understanding of how playing video games puts us into particular and unique affective relations to history, technology, and each other. Through affect theory, game studies gains an ethics and a politics rooted in an analysis of the collectivizing force of video games and the entrainment of our bodies through them. From game studies, affect theory gains a rigorous formal articulation of the ways certain shared and ordinary feelings take shape and reverberate through our everyday interaction with media and devices. Out of this convergence emerges a theoretical framework that addresses how our sense of being in the world, our feeling of the everyday, and our contemporary sensorium are shaped through our encounters with digital media. However, unlike Sicart, I argue that games *do* matter. Games are matter

The particular worlds, characters, and stories of video games cannot be meaningfully separated from what games ask us to do as players or from the less visible actions of their programming. Reconnecting play to everyday life does not require detaching it from the specific forms it takes through highly mediated objects. Thinking about video games as affective systems is not intended to replace play theory or proceduralism in game studies but rather to supplement them by addressing their omissions. Bringing affect theory into conversation with game studies accounts for the complex affective relationships that develop between players and video games, across bodies, code, molded plastic, and screens. Video games—as media objects, as cultural practices, and as structures of feeling—can tell us quite a bit about the collective desires, fears, and rhythms of everyday life in our precarious, networked, and procedurally generated world.

The collapsing of the computational metaphor—its ceasing to be a metaphor—is one of the backbones of structuralism, poststructuralism, and theories of affect. The wish to move away from the symbolic, away from representation, then occludes this structuring paradox: the very limitations of computational systems and their symbolic representations. It is easier in many ways to start fresh by proposing that computational media are not representations. This approach, however, leaves us wholly unable to address how representation still matters, and specifically how computational forms of representation increasingly matter. In holding affect theory and video games together, my approach assumes that video games—their images, sounds, stories, mechanics, and interfaces—can be read and interpreted as giving shape and form to particular feelings. Video games are not technologies capable of autonomously and unconsciously rewiring our bodies; rather, they are a particularly popular form of representation through which we can trace and analyze how affect moves across bodies and objects in the present. Furthermore, bodies and subjectivity have long been, or perhaps have always been, coassembled with technologies of mediation. Video games—like biometrics, affective computing, and motion capture—are an important contemporary site through which to address the longer history of how bodies are mediated, how their mediation informs shifting notions of what "the body" is, and which bodies come to matter. Thus, examining the links between affect and representation in video games is perhaps more crucial now than ever.

The cleaving apart of computation from representation that I trace in this book shapes not only how we analyze video games but also how these games enter into archival and curatorial spaces and discourses. In 2012, New York's Museum of Modern Art announced that it had begun acquiring video games for its permanent collection. Among the first games acquired were early arcade classics like *Asteroids, Space Invaders,* and *Pac-Man,* but also included were contemporary web-based virtual worlds such as *EVE Online* and *Minecraft.* MoMA's announcement aligned the institution with the wider movement to collect and preserve video games—from the Library of Congress and several prominent university archives in the United States to several more archives in the United Kingdom, Europe, and Japan. Paola Antonelli, MoMA's senior curator of architecture and design, explained the rationale for acquiring video games in terms of their significance to "interaction design." Echoing formalist approaches in game studies that cite action and proceduralism as essential to the medium, Antonelli stated, "Video games are the purest aspect of interaction design."[3] This, in turn, has shaped the ways MoMA has exhibited the acquired games. In a 2013–14 exhibition titled *Applied Design*, video games were displayed on screens recessed into the museum's walls, with their controllers perched in front of them on minimalist pedestals.[4] Gaming systems and arcade cabinets, it seems, were deemed too unsightly for display. Perhaps more to the point, these aspects of the material culture of games did not fit with the museum's definition of interaction design. The element of design that MoMA seeks to highlight through its collection and preservation is code, specifically source code. "What we want, what we aspire to, is the code," Antonelli explained, remarking on the museum's difficulty in acquiring code from game developers. "That is what would enable us to preserve the video games for a really long time."[5]

This faith in source code and the related downplaying of other material and immaterial aspects of video games is meant to address the problem of technological obsolescence. Institutions are faced with preserving software programs that were designed for platforms that become obsolete in a matter of a few years. Despite this hope that source code will equal preservation, scholars studying this problem often find just the opposite. Perfect source code, especially in the case of multiplayer online games, cannot preserve very much about what makes the game meaningful.[6] Similarly, in *Game After*, Raiford Guins argues for a richer appreciation for decay and fragments, and for the broken machines we can no longer turn on, in our understanding of what might constitute a video game's afterlife. He writes: "The museified state of video games suggests the need to reassess our definitions. . . . Treated as a 'cultural object,' video games will not 'come into being' if 'being' is premised exclusively on a machine's ability to be 'powered up' and have its software executed."[7] Following Guins, I would argue that a broken *Pac-Man* arcade cabinet, that which Antonelli dismisses as nostalgic arcade kitsch, can tell us more about the affective appeal of the game than the source code can.

All archives, of course, have a close relationship with decay and ruin. And archives are always imperfect and incomplete repositories of objects and ideas. It is impossible to create a perfect archival record of anything, let alone what a video game feels like, long after its technological moment has passed. But as interactive digital environments like video games become the ascendant medium of the twenty-first century, and as their formal and cultural properties continue to influence everything from art and cinema to how wars are fought and economic markets are run, it is becoming increasingly important to figure out how we will have access to and create meaning from the longer history of video games into the future. An affective archival practice for video games disrupts the fetishization of source code in game studies and game preservation. I am using the phrase *affective archive* to mean several things. As in Michel Foucault's sense of the archive, how might institutions that collect and preserve video games be more attendant to the myriad ways in which games, those who study them, and those who preserve them participate in a "system of discursivity" that is always necessarily incomplete yet establishes the possibility of what can be said about the games now and in the future?[8] In *An Archive of Feelings,* Ann Cvetkovich writes, "Cultural artifacts become the archive of something more ephemeral: culture as a 'way of life,' to borrow from Raymond Williams."[9] Video games, as I have argued in this book, are already affective archives. How might archival practices, then, address the ways video games are invested with feelings that "are encoded not only in the content of the texts themselves but in the practices that surround their production and reception"?[10]

The answers to these questions are outside the scope of this book, but I raise them in the spirit of thinking beyond the present state of game studies and toward what the field, specifically video game history, might look like in a few years. Bringing video games into conversation with affect theory depends, as this book illustrates, on a historicization of both terms, a process that is necessarily partial. The ways we tell video game history, and feel that history, in the future will be shaped by the partial affective archives that we form in the present. Indeed, the way the future will feel will also emerge from these ludic structures of feeling.

Acknowledgments

THE ACKNOWLEDGMENTS SECTION OF A BOOK provides insight into the web of connections and affective attachments that created the intellectual and emotional space for the work at hand. In a book about affect, I feel especially aware of this function. Each name here was typed with love and gratitude.

Many of the ideas developed in this book take the shape they do because of the people who taught me how to look, think, and write about media and art: Constance Penley, Christopher Newfield, Lisa Cartwright, Sharon Willis, Joan Saab, and Douglas Crimp.

A community of like-minded oddballs and badasses, whom I originally met in the Visual and Cultural Studies Program at the University of Rochester, has sustained my intellectual and emotional life into the present. Gloria Kim has been holding space (and other things) for me since I rescued her from being lost in Rush Rhees Library. Many others have been the crucial elements of a shared history and a shared project: thanks especially to Becky Burditt, Aviva Dove-Veibahn, Dinah Holtzman, Elizabeth Kalbfleisch, and Vicky Pass.

Half of this book was written in Toronto, where Nic Sammond provided space, in countless ways, for me to do the work. His emotional and intellectual generosity—and admirable work ethic—modeled for me a graceful and playful relationship to this strange life. This project never would have become a book proposal if not for a special writing group, also in Toronto: thank you, Dana Seitler, Elspeth Brown, Alla Gadasik, Eva-Lynn Jagoe, and Tess Takahashi.

I completed the manuscript in Ottawa with the help of an outstanding network of colleagues and friends: Rena Bivens, Mitchell Frank, Marc Furstenau, Christian Holz, Laura Horak, Gunnar Iversen, Franny Nudelman, Corrie Scott, Laura Taler, and Benjamin Woo. All, at many points, listened to my worries, fed me, lifted my spirits, and sent me back to work.

Ryan Conrad's indexing expertise saved me in the final exhausting stages of manuscript production.

The argument I lay out in this book benefited greatly from the thoughtful, careful, and generous advice of the manuscript's readers: Adrienne Shaw and Lisa Cartwright. What is good was made all the more so by their feedback. The errors, missteps, and failures remain all mine.

Finally, much love to my parents, Karen Anable-Nichols, Mark Nichols, Mark Anable, and Sally Anable, for loving me from a distance while I followed the long path toward this book.

Notes

Introduction

1. *E.T. the Extra-Terrestrial* was released for the Atari 2600 in late 1982. The game was a dismal failure and is somewhat apocryphally blamed for the North American video game crash of 1983–84. Electronic Arts was perhaps consciously foregrounding "art" and computers to distinguish the company's products from the struggling video game console market and Atari's poor-quality games.

2. As Linda Williams demonstrates, the "body genres"—films that evoke bodily responses, like weeping at melodramas or becoming sexually aroused by pornographic films—are generally excluded from the more "cerebral" art film canon. Linda Williams, "Film Bodies: Gender, Genre, and Excess," *Film Quarterly* 44, no. 4 (1991): 2–13.

3. Siegfried Kracauer, "Boredom," in *The Mass Ornament: Weimar Essays*, trans. and ed. Thomas Y. Levin (Cambridge, Mass.: Harvard University Press, 1995), 331–36.

4. Raymond Williams, "Culture Is Ordinary" (1958), in *The Everyday Life Reader*, ed. Ben Highmore (London: Routledge, 2001), 91.

5. Ibid., 93.

6. Ibid., 92.

7. Ibid., 93.

8. Patrick Jagoda, "Fabulously Procedural: Braid, Historical Processing, and the Videogame Sensorium," *American Literature* 85, no. 4 (2015): 768.

9. The intellectual response to cybernetics was to separate thoughts (as information) from feelings (the body). The separation of mind and the body, reason and the passions, and other thought/feeling distinctions predate cybernetics, of course, but in the mid-twentieth century computers and cybernetics provided a new take on this old theme.

10. Fred Turner traces the collapse of the computational metaphor from a different angle in *From Counterculture to Cyberculture: Stewart Brand, the Whole Earth Network, and the Rise of Digital Utopianism* (Chicago: University of Chicago Press, 2006).

11. Raymond Williams, "Structure of Feeling," in *Marxism and Literature* (Oxford: Oxford University Press, 1977), 128–35.

12. See, for example, Jane McGonigal's books *Reality Is Broken: Why Games Makes Us Better and How They Can Change the World* (New York: Penguin Press, 2011); and *Super Better: A Revolutionary Approach to Getting Stronger, Happier, Braver and More Resilient—Powered by the Science of Games* (New York: Penguin Press, 2015).

13. Miguel Sicart, *Play Matters* (Cambridge, Mass.: MIT Press, 2014), 1–6.

14. Adrienne Shaw, "Circles, Charmed and Magic: Queering Game Studies," *QED: A Journal in GLBTQ Worldmaking* 2, no. 2 (2015): 76.

15. See, for example, the articles in the inaugural issue of the online journal *Game Studies* 1, no. 1 (2001), http://gamestudies.org.

16. Ian Bogost, "Procedural Rhetoric," in *Persuasive Games: The Expressive Power of Videogames* (Cambridge, Mass.: MIT Press, 2007), 1–64.

17. Miguel Sicart, "Against Procedurality," *Game Studies* 11, no. 3 (2011), http://gamestudies.org; Brendan Keogh, "Across Worlds and Bodies: Criticism in the Age of Video Games," *Journal of Games Criticism* 1, no. 1 (2014): 1–26, http://gamescriticism.org.

18. Clara Fernández-Vara, *Introduction to Game Analysis* (New York: Routledge, 2014), 7.

19. Christopher A. Paul, *Wordplay and the Discourse of Video Games: Analyzing Words, Design, and Play* (New York: Routledge, 2012), 2.

20. Sara Ahmed, *Queer Phenomenology: Orientations, Object, Others* (Durham, N.C.: Duke University Press, 2006).

21. Jon Dovey and Helen W. Kennedy, "From Margin to Center: Biographies of Technicity and the Construction of Hegemonic Games Culture," in *The Players' Realm: Studies on the Culture of Video Games and Gaming*, ed. J. Patrick Williams and Jonas Heide Smith (Jefferson, N.C.: McFarland, 2007), 150–51.

22. Gregory Seigworth and Melissa Gregg identify no fewer than eight different approaches to affect theory, all with slightly to radically different interpretations, histories, theoretical frameworks, and implications for our understanding of the term. Gregory J. Seigworth and Melissa Gregg, "An Inventory of Shimmers," in *The Affect Theory Reader*, ed. Melissa Gregg and Gregory J. Seigworth (Durham, N.C.: Duke University Press, 2010), 6–8.

23. See, for example, Brian Massumi, *Parables for the Virtual: Movement, Affect, Sensation* (Durham, N.C.: Duke University Press, 2002); Erin Manning, *Politics of Touch: Sense, Movement, Sovereignty* (Minneapolis: University of Minnesota Press, 2007).

24. See Lisa Cartwright, *Moral Spectatorship: Technologies of Voice and Affect in Postwar Representations of the Child* (Durham, N.C.: Duke University Press, 2008); Richard Grusin, *Premediation: Affect and Mediality after 9/11* (London: Palgrave Macmillan, 2010).

25. Laine Nooney, "A Pedestal, a Table, a Love Letter: Archaeologies of Gender in Videogame History," *Game Studies* 13, no. 2 (2013), http://gamestudies.org.

1. Feeling History

1. *Colossal Cave Adventure* was originally written by William Crowther in the mid-1970s and was significantly revised and expanded by Don Woods for wider distribution through ARPANET later in the decade. See Rick Adams, "A History of 'Adventure,'" accessed August 7, 2017, http://rickadams.org/adventure/a_history .html.

2. Nick Montfort, *Twisty Little Passages: An Approach to Interactive Fiction* (Cambridge, Mass.: MIT Press, 2003), 85–97.

3. For an analysis of this history, see Jennifer S. Light, "When Computers Were Women," *Technology and Culture* 40, no. 3 (1999): 455–83.

4. Nooney, "A Pedestal, a Table, a Love Letter."

5. Ibid.

6. Ibid.

7. Eve Kosofsky Sedgwick and Adam Frank, "Shame in the Cybernetic Fold: Reading Silvan Tomkins," in *Shame and Its Sisters: A Silvan Tomkins Reader,* ed. Eve Kosofsky Sedgwick and Adam Frank (Durham, N.C.: Duke University Press, 1995), 1–28.

8. Nooney, "A Pedestal, a Table, a Love Letter," emphasis added.

9. Ibid., emphasis added.

10. Eugenie Brinkema, *The Forms of the Affects* (Durham, N.C.: Duke University Press, 2014), 21. Brinkema's claim that film theorists using affect have abandoned attention to form is hyperbolic and not really borne out in the work of her examples. Steven Shaviro and Lisa Cartwright, for example, using very different approaches to "affect," still pay a great deal of attention to close reading, visual analysis, and "form" in their work. Neither's work can be reduced to the "expressivity model" that Brinkema is critiquing.

11. Ibid., 36.

12. Gilles Deleuze, *The Fold: Leibniz and the Baroque,* trans. Tom Conley (Minneapolis: University of Minnesota Press, 1993).

13. Brinkema, *The Forms of the Affects,* 37.

14. Massumi, *Parables for the Virtual,* 5.

15. Ibid., 8.

16. Ruth Leys, "The Turn to Affect: A Critique," *Critical Inquiry* 37, no. 3 (2011): 434–72.

17. Sara Ahmed, "Happy Objects," in Gregg and Seigworth, *The Affect Theory Reader,* 29.

18. Sara Ahmed, "Orientations: Toward a Queer Phenomenology," *GLQ: A Journal of Lesbian and Gay Studies* 12, no. 4 (2006): 543.

19. Why people who play videos games do or do not identify as "gamers" is a complex and fraught topic. See, for example, Adrienne Shaw, "Do You Identify as a Gamer? Gender, Race, Sexuality, and Gamer Identity," *New Media & Society* 14, no. 1 (2012): 25–41.

20. Shaw, "Circles, Charmed and Magic," 65.

21. *Kentucky Route Zero* is episodic, and like contemporary television dramas, it is working through a long narrative arc. Cardboard Computer released the first of five acts in 2013, and at the time of this writing the fourth act had just been released.

22. For explanations and speculations about the game's many references, see Magnus Hildebrandt, "Kentucky Fried Zero, Parts 1, 2, and 3," Superlevel (blog), accessed August 7, 2017, https://superlevel.de.

23. Sedgwick and Frank, "Shame in the Cybernetic Fold," 12.

24. See, for example, N. Katherine Hayles, *How We Became Posthuman: Virtual Bodies in Cybernetics, Literature, and Informatics* (Chicago: University of Chicago Press, 1999); Turner, *From Counterculture to Cyberculture*; Eve Meltzer, *Systems We Have Loved: Conceptual Art, Affect, and the Antihumanist Turn* (Chicago: University of Chicago Press, 2013); Orit Halpern, *Beautiful Data: A History of Vision and Reason since 1945* (Durham, N.C.: Duke University Press, 2014).

25. N. Katherine Hayles, *My Mother Was a Computer: Digital Subjects and Literary Texts* (Chicago: University of Chicago Press, 2005), 3–4.

26. Homay King, *Virtual Memory: Time-Based Art and the Dream of Digitality* (Durham, N.C.: Duke University Press, 2015), 3.

27. Ibid., 6.

28. Turner, *From Counterculture to Cyberculture*, 11–39.

29. For a history of the Macy Conferences, see Steve Joshua Heims, *The Cybernetics Group* (Cambridge, Mass.: MIT Press, 1991).

30. Silvan Tomkins, *Affect, Imagery, Consciousness: The Complete Edition* (New York: Springer, 2008), 4:983.

31. Ibid., 1:64.

32. Ibid., 1:63.

33. Silvan S. Tomkins and Samuel Messick, eds., *Computer Simulation of Personality: Frontier of Psychological Theory* (New York: John Wiley, 1963).

34. Elizabeth A. Wilson, *Affect and Artificial Intelligence* (Seattle: University of Washington Press, 2010), 58–82.

35. Ibid., 63.

36. Ibid., 65.

37. Tomkins, *Affect, Imagery, Consciousness*, 1:4.

38. Sedgwick and Frank, "Shame in the Cybernetic Fold," 12.

39. Ibid., 13.

40. Alan Turing, "Computing Machinery and Intelligence," *Mind* 49 (1950): 433–60.

41. Jennifer Light, "Taking Games Seriously," *Technology and Culture* 49, no. 2 (2008): 348.

42. Cartwright, *Moral Spectatorship*; Grusin, *Premediation*.

43. Jagoda, "Fabulously Procedural," 748.

44. Montfort, *Twisty Little Passages*, 90; Terry Harpold, "Screw the Grue: Mediality, Metalepsis, Recapture," *Game Studies* 7, no. 1 (2007), http://gamestudies.org.

45. Krista Bonello Rutter Giappone, "Humor and Self-Reflexivity in Adventure Games," *Game Studies* 15, no. 1 (2015), http://gamestudies.org; Henri Bergson, *Laughter: An Essay on the Meaning of the Comic*, trans. Cloudesley Brereton and Fred Rothwell (New York: Macmillan, 1914).

46. The resonance between Chamberlain's domestic setting and Williams's kitchen table is not coincidental. Before *Kentucky Route Zero*, its creators made *A House in California*, a game that visually resembles *Mystery House* and requires the player to reflect on the ways ordinary domestic objects become affectively charged through the lens of memory.

47. Nooney, "A Pedestal, a Table, a Love Letter."

48. Wendy Hui Kyong Chun, "On 'Sourcery,' or Code as Fetish," *Configurations* 16 (2008): 323.

2. Touching Games

1. Superbrothers staff, "Inconsolable," Superbrothers HQ (blog), 2013, http://www.superbrothershq.com.

2. Maurice Merleau-Ponty, *Phenomenology of Perception*, trans. Colin Smith (London: Routledge and Kegan Paul, 1962), 92.

3. Sara Ahmed, *The Cultural Politics of Emotion* (New York: Routledge, 2004).

4. Laura U. Marks, *Touch: Sensuous Theory and Multisensory Media* (Minneapolis: University of Minnesota Press, 2002), 3.

5. Elizabeth Ellcessor, *Restricted Access: Media, Disability, and the Politics of Participation* (New York: New York University Press, 2016), 61–87.

6. Ibid., 63.

7. J. M. Graetz, "The Origin of *Spacewar!*," *Creative Computing* 7, no. 8 (1981), http://www.masswerk.at/spacewar/SpacewarOrigin.html.

8. Rita Raley, "Code.surface || Code.depth," *Dichtung Digital*, no. 36 (2006), http://www.dichtung-digital.de/en.

9. Steve Swink, *Game Feel: A Game Designer's Guide to Virtual Sensation* (Burlington, Mass.: Morgan Kaufman, 2009), 1–33.

10. Eugénie Shinkle, "Feel It, Don't Think: The Significance of Affect in the Study of Digital Games," in *Proceedings of DiGRA 2005 Conference: Changing Views—Worlds in Play* (Tampere, Finland: DiGRA, 2005), 2, emphasis added.

11. Valerie Walkerdine, *Children, Gender, and Video Games: Towards a Relational Approach to Multimedia* (London: Palgrave Macmillan, 2007), 25–26.

12. Eugénie Shinkle, "Video Games, Emotion and the Six Senses," *Media, Culture & Society* 30, no. 6 (2008): 907.

13. Mark Hansen, *New Philosophy for New Media* (Cambridge, Mass.: MIT Press, 2004), 23.

14. Walkerdine, *Children, Gender, and Video Games,* 26.

15. See "Gender Swap: The Experiment with The Machine to Be Another," The Machine To Be Another (blog), accessed August 7, 2017, http://www.themachinetobeanother.org.

16. Espen Aarseth, "Genre Trouble: Narrativism and the Art of Simulation," in *First Person: New Media as Story, Performance, and Game,* ed. Pat Harrigan and Noah Wardrip-Fruin (Cambridge, Mass.: MIT Press, 2004), 48, emphasis added.

17. Alexander R. Galloway, *Gaming: Essays on Algorithmic Culture* (Minneapolis: University of Minnesota Press, 2006), 2.

18. Alexander R. Galloway, *The Interface Effect* (Cambridge: Polity Press, 2012), 63.

19. Ibid., 54.

20. Ian Bogost, *Unit Operations: An Approach to Videogame Criticism* (Cambridge, Mass.: MIT Press, 2006), 27, emphasis added.

21. Ibid., 98.

22. See, for example, Sicart "Against Procedurality"; Fernández-Vara, *Introduction to Game Analysis*; Paul, *Wordplay and the Discourse of Video Games*; Keogh, "Across Worlds and Bodies."

23. Frans Mäyrä, *An Introduction to Game Studies: Games in Culture* (London: Sage, 2008), 2, 17–18.

24. Ibid., 16.

25. Laura Mulvey, "Visual Pleasure and Narrative Cinema," *Screen* 16, no. 4 (1975): 6–18.

26. Massumi, *Parables for the Virtual,* 1–2.

27. Ibid., 14.

28. Ibid., 25.

29. Leys, "The Turn to Affect," 450–51.

30. Galloway, *The Interface Effect,* 128–29.

31. Ahmed, *The Cultural Politics of Emotion,* 1; Ahmed, "Orientations," 543.

32. Ahmed, *The Cultural Politics of Emotion,* 10.

33. Sedgwick and Frank, "Shame in the Cybernetic Fold," 14.

34. Karen Barad, "Quantum Entanglements and Hauntalogical Relations of Inheritance: Dis/continuities, SpaceTime Enfoldings, and Justice to Come," *Derrida Today* 3, no. 2 (2010): 240–68; Karen Barad, "On Touching—The Inhuman That I Therefore Am (V.1.1)," in *Power of Material/Politics of Materiality,* ed. Susanne Witzgall and Kerstin Stakemeier (Zurich: diaphanes, 2014), 153–64, revision of the version first published in *differences* 23, no. 3 (2012): 206–23.

35. Silvan Tomkins, "Simulation of Personality: The Interrelations between Af-

fect, Memory, Thinking, Perception, and Action," in Tomkins and Messick, *Computer Simulation of Personality*, 18.

36. Wilson, *Affect and Artificial Intelligence*.

37. Tomkins, "Simulation of Personality," 23.

38. Ibid., 20–21.

39. Silvan Tomkins, "Silvan Tomkins—The Skin Is Where It's At," plenary address to the International Society for Research and Emotion, Princeton University, 1990, transcription from video, https://www.youtube.com/watch?v=JkvJ_xuhRJc.

40. Tomkins, "Simulation of Personality," 22.

41. Jacques Derrida, *On Touching—Jean-Luc Nancy*, trans. Christine Irizarry (Stanford, Calif.: Stanford University Press, 2005), 75.

42. Barad, "On Touching," 3.

43. Ibid., 8.

3. Rhythms of Work and Play

1. These figures come from documents filed with the Securities and Exchange Commission by King, Inc., at the time of its initial public offering, February 18, 2014, http://www.sec.gov.

2. Sam Anderson, "Just One More Game . . . Angry Birds, Farmville and Other Hyperaddictive 'Stupid Games,'" *New York Times Magazine*, April 4, 2012.

3. See Johan Huizinga, *Homo Ludens: A Study of the Play-Element in Culture* (1938; repr., Boston: Beacon Press, 1971), 28–29; Sicart, *Play Matters*, 1–18.

4. Sianne Ngai, *Our Aesthetic Categories: Zany, Cute, Interesting* (Cambridge, Mass.: Harvard University Press, 2012), 183.

5. Jesper Juul, *A Casual Revolution: Reinventing Video Games and Their Players* (Cambridge, Mass.: MIT Press, 2012), 37.

6. "Survey: One-in-Four White-Collar Gamers Play at Work—Senior Executives Have Most Fun," PR Newswire, September 4, 2007, http://www.prnewswire.co.uk.

7. Michel de Certeau, *The Practice of Everyday Life*, trans. Steven Randall (Berkeley: University of California Press, 1984), 25.

8. For an example of this argument, see Nick Dyer-Witheford and Greig de Peuter, *Games of Empire: Global Capitalism and Video Games* (Minneapolis: University of Minnesota Press, 2009), 27–28.

9. Stewart Brand, "Spacewar: Fanatic Life and Symbolic Death among the Computer Bums," *Rolling Stone*, December 7, 1972.

10. Turner, *From Counterculture to Cyberculture*, 24, 116.

11. Maurizio Lazzarato, "Immaterial Labor," in *Radical Thought in Italy: A Potential Politics*, ed. Paolo Virno and Michael Hardt (Minneapolis: University of Minnesota Press, 1996), 133.

12. Michael Hardt, "Affective Labor," *boundary 2* 26, no. 2 (1999), 97.

13. Jonathan Crary discusses this kind of time in *24/7: Late Capitalism and the Ends of Sleep* (London: Verso, 2013).

14. Melissa Gregg, *Work's Intimacy* (Cambridge: Polity Press, 2011), 18.

15. For discussion of such services, see Alison Hearn, "Structuring Feeling: Web 2.0, Online Ranking and Rating, and the Digital 'Reputation' Economy," *Ephemera* 10, nos. 3–4 (2010): 421–38.

16. Sarah Banet-Weiser, "Branding the Post-feminist Self: Girls' Video Production and YouTube," in *Mediated Girlhoods: New Explorations of Girls' Media Culture,* ed. Mary Celeste Kearney (New York: Peter Lang, 2011), 277–93.

17. Mark Andrejevic, "Privacy, Exploitation, and the Digital Enclosure," *Amsterdam Law Forum* 1, no. 4 (2009): 50, http://amsterdamlawforum.org.

18. Christel Schoger, "Publication: 2013 Year in Review," *Distimo,* December 17, 2013.

19. Shift Worker (Production Shed, 2013), iOS; Procraster (Simon Sorboe Solbakken, 2013), iOS; Life Graphy (ByungWook Kang, 2014), iOS; OmniFocus (OmniGroup, 2013), iOS; 24me (24me Ltd, 2013), iOS; Grid (Binary Thumb, 2013), iOS.

20. Shira Chess, "A Time for Play: Interstitial Time, Invest/Express Games, and Feminine Leisure Style," *New Media & Society* (published online July 28, 2016): 7, doi:10.1177/1461444816660729.

21. Casual Games Association, "Casual Games Market Report 2007," "Social Network Games: Casual Games Sector Report 2012," and "Mobile Gaming: Casual Games Sector Report 2012," http://www.cga.global.

22. John Vanderhoef, "Casual Threats: The Feminization of Casual Video Games," *Ada: A Journal of Gender, New Media, and Technology,* no. 2 (2013), http://adanewmedia.org.

23. Among the notable exceptions are Jesper Juul, John Vanderhoef, and Mia Consalvo, "Using Your Friends 2.0: Social Mechanics in Social Games," in *FDG 2011: Proceedings of the 6th International Conference on Foundations of Digital Games* (New York: ACM, 2011), 188–95; Shira Chess, "Going with the Flo: *Diner Dash* and Feminism," *Feminist Media Studies* 12, no. 1 (2012): 83–99.

24. Ian Bogost, *How to Do Things with Videogames* (Minneapolis: University of Minnesota Press, 2011), 83.

25. Ibid., 87.

26. Andreas Huyssen lays out this history well in *After the Great Divide: Modernism, Mass Culture, Postmodernism* (Bloomington: Indiana University Press, 1986), 44–64.

27. Chess, "Going with the Flo," 90.

28. Ibid.; Chess cites Arlie Russell Hochschild, *The Time Bind* (New York: Holt, 2001).

29. Chess, "Going with the Flo," 91–92.

30. Mihaly Csikszentmihalyi, *Beyond Boredom and Anxiety: Experiencing Flow in Work and Play* (San Francisco: Jossey-Bass, 1975). For a recent example of the

application of Csikszentmihalyi's theory to video games, see Katherine Isbister, *How Games Move Us: Emotion by Design* (Cambridge, Mass.: MIT Press, 2016).

31. Braxton Soderman, "Intrinsic Motivation: Flow, Video Games, and Participatory Culture," *Transformative Works and Cultures* 2 (March 2009): para. 3.4, http://journal.transformativeworks.org.

32. Juul, *A Casual Revolution*, 68.

33. Cathy Davidson, "So Last Century," *Times Higher Education*, April 28, 2011, https://www.timeshighereducation.com.

34. Ngai, *Our Aesthetic Categories*, 9.

35. Ibid., 10, emphasis added; Ngai cites Nikolas Rose, *Governing the Soul: The Shaping of the Private Self* (London: Free Association Books, 1999), 70–71.

36. It is tempting to say that this rhythm of pauses and breaks is found primarily in what used to be called "white-collar" labor. But between the labor shifts that have made once-stable white-collar jobs more precarious, the sheer variety of work that involves sitting in front of a computer, and the ubiquity of mobile phones with games on them, the rhythm I am talking about here extends beyond the narrow category of white-collar office worker. Games are in all kinds of workplaces, from a call center in Mumbai to a nurses' break room in Toronto to a stockbroker's office in Manhattan.

37. Tania Modleski, "Rhythms of Reception: Daytime Television and Women's Work," in *Regarding Television: Critical Approaches*, ed. E. Ann Kaplan (Los Angeles: University Publications of America, 1983), 71–74.

38. Chess, "Going with the Flo," 96; Chess cites Arlie Russell Hochschild, *The Managed Heart: Commercialization of Human Feeling* (Berkeley: University of California Press, 1983).

39. Ngai, *Our Aesthetic Categories*, 8.

40. Ibid., 188.

41. Massumi, *Parables for the Virtual*, 26.

42. *Diner Dash* Facebook page, "Status Update," March 11, 2011, 5:21 p.m.

43. See Lauren Berlant, *The Female Complaint: The Unfinished Business of Sentimentality in American Culture* (Durham, N.C.: Duke University Press, 2008); Lauren Berlant, *Intimacy* (Chicago: University of Chicago Press, 2000); Lauren Berlant, *The Queen of America Goes to Washington City* (Durham, N.C.: Duke University Press, 1997).

44. Berlant, *The Female Complaint*, 2.

45. Ibid., 1–2.

46. Ibid., 3.

4. Games to Fail With

1. Huizinga, *Homo Ludens*, 6.

2. Lauren Berlant, *Cruel Optimism* (Durham, N.C.: Duke University Press, 2011), 4.

3. Ibid., 7.

4. Andrea Scott, "Futurism: Cory Arcangel Plays around with Technology," *New Yorker,* May 30, 2011, http://www.newyorker.com.

5. Berlant, *Cruel Optimism,* 2.

6. *Oxford English Dictionary Online,* s.v. "failure," accessed August 8, 2017, http://www.oed.com.

7. Scott A. Sandage, *Born Losers: A History of Failure in America* (Cambridge, Mass.: Harvard University Press, 2006), 4–5.

8. Tomkins, *Affect, Imagery, Consciousness,* 2:359.

9. Ibid., 2:361, emphasis added.

10. Ibid., 2:388.

11. Jesper Juul, *The Art of Failure: An Essay on the Pain of Playing Video Games* (Cambridge, Mass.: MIT Press, 2013), 1–31.

12. Roger Caillois, *Man, Play, and Games,* trans. Meyer Barash (1961; repr., Urbana: University of Illinois Press, 2001), 14–19.

13. Pippin Barr, *You Say Jump I Say How High,* accessed August 8, 2017, http://www.pippinbarr.com.

14. Pippin Barr, *Don't Drown,* accessed August 8, 2017, http://www.pippinbarr.com.

15. Pippin Barr, *Let's Play: Ancient Greek Punishment,* accessed August 8, 2017, http://www.pippinbarr.com.

16. Pippin Barr, "Seduction by iPhone" (blog post), June 4, 2013, n.p., http://www.pippinbarr.com.

17. Ibid.

18. Berlant, *Cruel Optimism,* 3.

19. Ibid., 198.

20. Ngai, *Our Aesthetic Categories,* 2.

21. Ed Halter, quoted in Blythe Sheldon, "Graphic Violence: Mark Essen's Brutal, Lo-Fi Video Games Are about to Make Him an Art-World Star," *New York Magazine,* March 29, 2009, http://nymag.com.

22. Tomkins, *Affect, Imagery, Consciousness,* 2:361.

23. Judith (Jack) Halberstam, *The Queer Art of Failure* (Durham, N.C.: Duke University Press, 2011), 3.

24. Jamie Peck, *Constructions of Neoliberal Reason* (Oxford: Oxford University Press, 2010).

25. See, for example, Galloway, *Gaming;* Bogost, *Unit Operations;* and three articles in *Game Studies* 1, no. 1 (2001), http://gamestudies.org: Espen Aarseth, "Computer Game Studies, Year One"; Jesper Juul, "Games Telling Stories?"; and Markku Eskelinen, "The Gaming Situation."

26. Graeme Kirkpatrick, *Aesthetic Theory and the Video Game* (Manchester: Manchester University Press, 2011), 13.

27. Ibid., 1–2, emphasis added.

28. The separation of form from content, or representation from computation, in game studies mirrors in effect the rigid separation of affect from emotion in some accounts of affect theory. As discussed elsewhere in this book, some versions of affect theory grounded in neurobiology tend to evacuate ideology from representation. This is part of Ruth Ley's critique of Brian Massumi in her essay "The Turn to Affect." In these models, it can seem as though it is no longer necessary to analyze the content of films, novels, or video games because they act on us in precognitive and presubjective ways. This is also how affect sometimes gets taken up in new materialism and object-oriented ontology, where the real interest in cultural objects resides not in the field of representation but in the objects' positions in a flattened ontology of being in and among a network of objects. See, for example, Jane Bennett, *Vibrant Matter: A Political Ecology of Things* (Durham, N.C.: Duke University Press, 2010); Ian Bogost, *Alien Phenomenology, or What It's Like to Be a Thing* (Minneapolis: University of Minnesota Press, 2012).

29. Kirkpatrick, *Aesthetic Theory and the Video Game,* 3–4.

30. See, for example, Bruce Sterling, "An Essay on the New Aesthetic," *Wired,* April 2012, https://www.wired.com.

31. Jodi Dean, *Democracy and Other Neoliberal Fantasies: Communicative Capitalism and Left Politics* (Durham, N.C.: Duke University Press, 2009).

32. For an analysis of failure and humor in fumblecore games, see Ian Bryce Jones, "Do the Locomotion: Obstinate Avatars, Dehiscent Performances, and the Rise of the Comedic Video Game," *Velvet Light Trap,* no. 77 (Spring 2016): 86–99.

33. Bergson, *Laughter,* 1–22.

34. Ngai, *Our Aesthetic Categories,* 174.

35. Bergson, *Laughter,* 32.

36. Cory Arcangel's Official Portfolio Website and Portal, accessed August 8, 2017, http://www.coryarcangel.com.

37. Ngai, *Our Aesthetic Categories,* 4.

38. Walter Benjamin, "The Work of Art in the Age of Its Technological Reproducibility (Second Version)," in *The Work of Art in the Age of Its Technological Reproducibility, and Other Writings on Media,* ed. Michael W. Jennings, Brigid Doherty, and Thomas Y. Levin, trans. Edmund Jephcott, Rodney Livingstone, Howard Eiland, et al. (Cambridge, Mass.: Belknap Press, 2008), 26.

39. Miriam Bratu Hansen, "Room-for-Play: Benjamin's Gamble with Cinema," *October* 109 (2004): 6, emphasis added.

40. Ibid., 8.

41. Peter Krapp, *Noise Channels: Glitch and Error in Digital Culture* (Minneapolis: University of Minnesota Press, 2011), 76.

42. Ibid., 75–92.

Conclusion

1. Sicart, *Play Matters*, 2.

2. Ibid., 18.

3. Paola Antonelli, "Why I Brought Pac-Man to MoMA," TED Salon New York City, May 2013, video, https://www.ted.com.

4. *Applied Design*, organized by Paola Antonelli, senior curator, and Kate Carmody, curatorial assistant, Department of Architecture and Design, Museum of Modern Art, New York, March 2, 2013–January 20, 2014.

5. Antonelli, "Why I Brought Pac-Man to MoMA."

6. Jerome McDonough, Robert Olendorf, Matthew Kirschenbaum, Kari Kraus, Doug Reside, Rachel Donahue, Andrew Phelps, Christopher Egert, Henry Lowood, and Susan Rojo, *Preserving Virtual Worlds: Final Report* (Urbana-Champaign: Graduate School of Library and Information Science, University of Illinois, 2010), http://hdl.handle.net/2142/17097.

7. Raiford Guins, *Game After: A Cultural Study of Video Game Afterlife* (Cambridge, Mass.: MIT Press, 2014), 32.

8. Michel Foucault, *The Archaeology of Knowledge and The Discourse on Language*, trans. A. M. Sheridan Smith (New York: Pantheon Books, 1972), 129.

9. Ann Cvetkovich, *An Archive of Feelings: Trauma, Sexuality, and Lesbian Public Cultures* (Durham, N.C.: Duke University Press, 2003), 9.

10. Ibid., 7.

Index

Dean, Jodi, 123
de Certeau, Michel, 75
Deleuze, Gilles, xi, 6
Derrida, Jacques, 64
Diner Dash, 74, 90, 93, 95; description, 83, 85–88; social networks, 98
Don't Drown, 111
Dovey, Jon, xvi
Duck Hunt, 10
Durations, 111
Dys4ia, 38, 56–58

Electronic Arts, 137n1; marketing, vii, ix, x, xi
Ellcessor, Elizabeth, 41
Empathy Machine, The, 38, 46–48, 56, 58
E.T. the Extra Terrestrial, 137n1
EVE Online, 132

failure: aesthetics, 119–24, 127–29; definition, 107; personalization, 107, 109; uses, 104–5
FarmVille, 72, 74; social networks, 98
feminism: affective labor, 92–93; engagement with video games, 16–17, 33, 122; game studies and, 2, 5, 16, 40; interventions, xvi, xvii, xx; theory, xviii, 8–9, 59, 93; video game culture and, xvii, 121; video game history, xx, 3
Fernández-Vara, Clara, xv
Foucault, Michel, 134
Frank, Adam, 4, 17, 21–22, 59

Galaga, 10
Galloway, Alexander, 49–50, 52, 55–56
gamergate, xvii, 35
game studies: affect theory and, 131; antirepresentational turn, 49–50,

69; challenges in field, 67–68; computational approach, 51–52, 55–56; critiques, xv, 84, 120–21; debates within, 35–36; feminist approaches, 2–3, 122; future, xv, 36, 131, 134; gender and, 56, 69, 82, 84, 122; ludological approach, 48; origin, 35; sensory deficit, 47
gamification, xiii, 76
gender, 100; casual games, 82–84, 100–101; embodied experience, 47, 56; game studies and, 56, 69, 82, 84, 122; gaming culture and, xvi, xvii, 52, 121; identity, 47; inequality, 99; as infrastructure, 6, 8–9, 23, 31–34; *Kentucky Route Zero*, 12, 29; labor, 93; player demographics, 10–12, 121; spatial imaginary, 52; transgender identity, 47; video game history, 3; video games and, 35, 44, 46, 57, 66. *See also* feminism: affect and
Goeller, Kelly, 123
Gregg, Melissa, 77
Grimonprez, Johan, 123
Guins, Raiford, 133

Halberstam, Jack, 119–20
Halo, 103
Halt and Catch Fire, 34–35
Halter, Ed, 117
Hansen, Mark, 45
Hansen, Miriam, 128
haptic visuality, 40–41
Hardt, Michael, 76
Hayles, N. Katherine, 18
Higinbotham, William, x
Huizinga, Johan, 104, 131
human–computer interaction (HCI), 41–43, 60
humor, 25–27, 29, 103, 124–25

Aubrey Anable is assistant professor of film studies at Carleton University.

Milton Keynes UK
Ingram Content Group UK Ltd.
UKHW022156141223
434411UK00014B/379